Data Security in Internet of Things Based RFID and WSN Systems Applications

Internet of Everything (IoE)

Series Editor:
Vijender Kumar Solanki, Raghvendra Kumar, and Le Hoang Son

IOT
Security and Privacy Paradigm
Edited by Souvik Pal, Vicente Garcia Diaz, and Dac-Nhuong Le

Smart Innovation of Web of Things
Edited by Vijender Kumar Solanki, Raghvendra Kumar, and Le Hoang Son

Big Data, IoT, and Machine Learning
Tools and Applications
Rashmi Agrawal, Marcin Paprzycki, and Neha Gupta

Internet of Everything and Big Data
Major Challenges in Smart Cities
*Edited by Salah-ddine Krit, Mohamed Elhoseny, Valentina Emilia Balas,
Rachid Benlamri, and Marius M. Balas*

Bitcoin and Blockchain
History and Current Applications
*Edited by Sandeep Kumar Panda, Ahmed A. Elngar, Valentina Emilia Balas,
and Mohammed Kayed*

Privacy Vulnerabilities and Data Security Challenges in the IoT
Edited by Shivani Agarwal, Sandhya Makkar, and Tran Duc Tan

Handbook of IoT and Blockchain
Methods, Solutions, and Recent Advancements
*Edited by Brojo Kishore Mishra, Sanjay Kumar Kuanar, Sheng-Lung Peng,
and Daniel D. Dasig, Jr.*

Blockchain Technology
Fundamentals, Applications, and Case Studies
Edited by E Golden Julie, J. Jesu Vedha Nayahi, and Noor Zaman Jhanjhi

Data Security in Internet of Things Based RFID and
WSN Systems Applications
Edited by Rohit Sharma, Rajendra Prasad Mahapatra, and Korhan Cengiz

For more information about this series, please visit: https://www.crcpress.com/
Internet-of-Everything-IoE-Security-and-Privacy-Paradigm/book-series/CRCIOESPP

Data Security in Internet of Things Based RFID and WSN Systems Applications

Edited by
Rohit Sharma, Rajendra Prasad Mahapatra, and
Korhan Cengiz

CRC Press
Taylor & Francis Group
Boca Raton London New York

CRC Press is an imprint of the
Taylor & Francis Group, an **informa** business

First edition published 2021
by CRC Press
6000 Broken Sound Parkway NW, Suite 300, Boca Raton, FL 33487-2742

and by CRC Press
2 Park Square, Milton Park, Abingdon, Oxon, OX14 4RN

© 2021 Taylor & Francis Group, LLC

CRC Press is an imprint of Taylor & Francis Group, LLC

Library of Congress Cataloging in Publication Data
Names: Sharma, Rohit, editor. | Mahapatra, Rajendra Prasad, editor. |
Cengiz, Korhan, editor.
Title: Data security in internet of things based RFID and WSN systems
applications / edited by Rohit Sharma, Rajendra Prasad Mahapatra and
Korhan Cengiz.
Description: Boca Raton : CRC Press, 2020. | Series: Internet of everything
(ioe): security and privacy paradigm | Includes bibliographical references and index. |
Summary: "This book will focus on RFID (Radio Frequency Identification),
IoT (Internet of Things), and WSN (Wireless Sensor Network). It will include contributions
that discuss the security and privacy issues as well as the opportunities and applications
that are tightly linked to sensitive infrastructures and strategic services. This book will address
the complete functional framework and workflow in IoT-enabled RFID systems and explore the
basic and high-level concepts. It is based on the latest technologies, and covers the major challenges,
issues, and advances in the field. It will present data acquisition and case studies related to data
intensive technologies in RFID-based IoT and includes WSN based systems and its security.
It can serve as a manual for those in the industry while also helping beginners to understand both
the basic and the advanced aspects in IoT-based RFID related issues. The book can be a
premier interdisciplinary platform for researchers, practitioners, and educators to present and
discuss the most recent innovations, trends, and concerns as well as practical challenges encountered
and find solutions that have been adopted in the fields of IoT and analytics"—Provided by publisher.
Identifiers: LCCN 2020026421 (print) | LCCN 2020026422 (ebook) |
ISBN 9780367260439 (hardback) | ISBN 9780429294990 (ebook)
Subjects: LCSH: Radio frequency identification systems—Security measures. |
Wireless sensor networks—Security measures. | Internet of things—Security measures.
Classification: LCC TK6570.I34 D38 2020 (print) | LCC TK6570.I34 (ebook) |
DDC 006.2/45028558—dc23

ISBN: 978-0-367-26043-9 (hbk)
ISBN: 978-0-429-29499-0 (ebk)

Typeset in Times
by codeMantra

Contents

Preface

This edited book aims to bring together leading academic scientists, researchers, and research scholars to exchange and share their experiences and research results on all aspects of wireless IoT and analytics.

It also provides a premier interdisciplinary platform for researchers, practitioners, and educators to present and discuss the most recent innovations, trends, and concerns, as well as practical challenges encountered and solutions adopted in the fields of IoT and analytics.

ORGANIZATION OF THE BOOK

This book is organized into nine chapters. A brief description of each of the chapters is as follows:

Chapter 1 discusses the estimation of "An Intelligent and Optimistic Disease Diagnosis." In this chapter, the authors discuss the IoT-based smart city paradigm.

Chapter 2 focuses on the major aspects of soft computing techniques for Boolean function and reliability-based approach of blood bank supply chain management. This chapter targets the design of vector-evaluated genetic algorithm.

Chapter 3 deals with the design of energy- and power-efficient IoT-enabled smart park.

Chapter 4 presents the use of Raspberry Pi IoT foundation features of the representation-driven advancement process for machine learning home security system.

Chapter 5 analyzes the investigation of 48 leading ERP systems sold on the Polish market to answer this question.

Chapter 6 presents a new approach to achieve high extinction ratio (ER) at low $V_\pi L$ with good bit error rate (BER) for high data rate transmission.

Chapter 7 deals with blood bank storage inventory models that have been developed to account for the effects of decay of goods with ramp-type demand and inflation using particle swarm optimization.

Chapter 8 highlights the virtualization, energy efficiency, virtual machine (VM) migration mechanism, and types of live VM migration techniques.

Chapter 9 discusses the dynamic exchange buffer switching and blocking control in wireless sensor networks.

Dr Rohit Sharma
Dr Rajendra Prasad Mahapatra
Dr Korhan Cengiz

Editors

Rohit Sharma is an Assistant Professor in the Department of Electronics and Communication Engineering, SRM Institute of Science and Technology, Delhi NCR Campus Ghaziabad, India. He is an active member of ISTE, IEEE, ICS, IAENG, and IACSIT. He is an editorial board member and reviewer of more than 12 international journals and conferences, including the topmost journal *IEEE Access* and *IEEE Internet of Things Journal*. He serves as a Book Editor for 7 different titles to be published by CRC Press, Taylor & Francis Group, USA, and Apple Academic Press, CRC Press, Taylor & Francis Group, USA, Springer, etc. He has received the Young Researcher Award in "2nd Global Outreach Research and Education Summit & Awards 2019" hosted by Global Outreach Research & Education Association (GOREA). He is serving as Guest Editor in *SCI Journal* of Elsevier. He has actively been an organizing end of various reputed International conferences. He is serving as an Editor and Organizing Chair to 3rd Springer International Conference on Microelectronics and Telecommunication (2019), and have served as the Editor and Organizing Chair to 2nd IEEE International Conference on Microelectronics and Telecommunication (2018), Editor and Organizing Chair to IEEE International Conference on Microelectronics and Telecommunication (ICMETE-2016) held in India, Technical Committee member in "CSMA2017, Wuhan, Hubei, China," "EEWC 2017, Tianjin, China," IWMSE2017 "Guangzhou, Guangdong, China," "ICG2016, Guangzhou, Guangdong, China," and "ICCEIS2016 Dalian Liaoning Province, China."

Rajendra Prasad Mahapatra, BTech, MTech, PhD, is currently working at SRM Institute of Science & Technology, NCR Campus, as Professor and Head – Computer Science and Engineering and Dean Admissions. Prof. Mahapatra has vast experience of 17 years as an academician, researcher, and administrator. During these 17 years, he has worked in India and abroad. He has been associated with Mekelle University, Ethiopia, for more than 2 years. Two candidates have successfully completed their PhD under his supervision, and seven students are perusing their research under his guidance. Prof. Mahapatra has authored more than 70 research papers, which are published in international journals of publishers such as Inderscience, Emerald, Elsevier, IEEE, and Springer. He is a fellow member of I.E. (India), senior member of IACSIT (Singapore), life member of ISTE, member of IEEE, and many more reputed bodies.

Korhan Cengiz received his BS degree in Electronics and Communication Engineering in 2008 from Kocaeli University, Turkey. He received his PhD degree in Electronics Engineering in 2016 from Kadir Has University, Turkey. Dr. Cengiz has more than 30 articles related to wireless sensor networks and wireless communications. Dr. Cengiz serves as a TPC member for more than 20 conferences. He is editor-in-chief of two journals and editor of several journals. His honors include Tubitak Priority Areas PhD scholarship and two best paper awards in conferences ICAT 2016 and ICAT 2018.

Contributors

Aditya Agarwal
Department of Electronics and
   Communication Engineering
SRM Institute of Science & Technology
Ghaziabad, Uttar Pradesh, India

Navin Ahlawat
Department of Computer Science
SRM Institute of Science & Technology
Delhi-NCR Campus
Ghaziabad, Uttar Pradesh, India

Dhowmya Bhatt
Department of Information Technology
SRM Institute of Science & Technology
Delhi-NCR Campus
Ghaziabad, Uttar Pradesh, India

Ludosław Drelichowski
The University of Economy
Bydgoszcz, Poland

R. Ganesh Babu
Department of Electronics and
   Communication Engineering
SRM TRP Engineering College
Tiruchirappalli, Tamilnadu, India

Saptarshi Gupta
Department of Electronics and
   Communication Engineering
SRM Institute of Science & Technology
Ghaziabad, Uttar Pradesh, India

R. Jayakumar
Department of Electronics and
   Instrumentation Engineering
Erode Sengunthar Engineering College
Erode, Tamilnadu, India

R. G. Jesuwanth Sugesh
School of Electronics Engineering
Vellore Institute of Technology
Chennai, Tamilnadu, India

P. Karthika
Department of Computer Applications
Kalasalingam Academy of Research
   and Education
Krishnan Koil, Tamilnadu, India

Praveen Kumar Malik
Department of Electronics and
   Communication Engineering
Lovely Professional University
Phagwara, Punjab, India

Tripti Pandey
Department of Computer Science and
   Engineering
SRM Institute of Science &
   Technology
Delhi-NCR Campus
Ghaziabad, Uttar Pradesh, India

Zdzislaw Polkowski
Jan Wyzykowski University
Polkowice, Poland

Satya Sai Srikant
Department of Electronics and
   Communication Engineering
SRM Institute of Science & Technology
Ghaziabad, Uttar Pradesh, India

M. N. Saravana Kumar
Department of Electronics and
   Instrumentation Engineering
Erode Sengunthar Engineering College
Erode, Tamilnadu, India

Manash Sarkar
Department of Computer Science and
   Engineering
SRM Institute of Science & Technology
Ghaziabad, Uttar Pradesh, India

Anurag Singh
Department of Electronics and
   Communication Engineering
SRM Institute of Science & Technology
Ghaziabad, Uttar Pradesh, India

Dinesh Singh
Department of Computer Science and
   Engineering
Deenbandhu Chhotu Ram University of
   Science and Technology
Murthal, Sonepat, Haryana, India

Shalu Singh
Department of Computer Science and
   Engineering
Deenbandhu Chhotu Ram University of
   Science and Technology
Murthal, Sonepat, Haryana, India

Ajay Singh Yadav
Department of Mathematics
SRM Institute of Science & Technology
Delhi-NCR Campus
Ghaziabad, Uttar Pradesh, India

A. Sivasubramanian
School of Electronics Engineering
Vellore Institute of Technology
Chennai, Tamilnadu, India

Anupam Swami
Department of Mathematics
Government Post Graduate College
Sambhal, Uttar Pradesh, India

Sławomir Świtek
The University of Economy
Bydgoszcz, Poland

P. Vidhya Saraswathi
Department of Computer Applications
Kalasalingam Academy of Research
   and Education
Krishnan Koil, Tamilnadu, India

1 An Intelligent and Optimistic Disease Diagnosis

An IoT-Based Smart City Paradigm

Manash Sarkar, Saptarshi Gupta, Satya Sai Srikant, Anurag Singh and Aditya Agarwal
SRM Institute of Science & Technology

CONTENTS

1.1   INTRODUCTION

The enormous deployment of Internet of Things (IoT) is permitting smart city ventures and activities everywhere throughout the world. The IoT is a secluded way to deal with combined different sensors with all the ICT arrangements. More than 50 billion items will be associated and conveyed in savvy urban communities in 2020. The core of savvy urban community's activities is the IoT interchanges. IoT is intended to help smart city idea, which targets using the most progressive correspondence advances to advance administrations for the organization of the city and the residents. Since the past few decades, a virtual revolution has been growing rapidly with the use of simulation technology for clinical functions technologically. It becomes advanced within the computational area, speed and power, graphics and image rendering, show systems, tracking, interface technology, tactile devices, and authoring package. Artificial intelligence (AI) has supported the creation of affordable and usable applications-based systems.

    The IoT is an arrangement of interrelated figuring gadgets, mechanical and computerized machines, items, creatures, or individuals who are given special identifiers (unique identifiers [UIDs]) and the capacity to move information over a system without expecting human-to-PC cooperation or human-to-human communication [1–5]. According to the International Telecommunication Union, "The IoT can be viewed as a global infrastructure for the information society, enabling advanced services by interconnecting (physical and virtual) things based on existing and evolving interoperable information and communication technologies (ICT)." Thing: With respect to the IoT, this is an object of the physical world (physical things) or the data world (virtual things), which is fit for being recognized and incorporated into correspondence systems [6]. IoT gadget can be used to control the various equipments used in building. We can monitor performance of electrical equipments remotely, and it can provide ON/OFF facility remotely. Power meter of the smart building can be integrated with the IoT, which can provide less manpower to monitor the readings which results in smart billing of power consumption in smart building [7,8]. Applications of this type consist of sensors and mobile app for sketch of map so that a user can search the area temperature, map, locations of different place, environmental monitoring, and so on [9]. IoT can be applied for environmental monitoring such as weather monitoring, wind speed, humidity, rain water measurement, and so on [10]. It can also be used for forest monitoring such as wild animal and wildlife monitoring [11–13], protection of crop from animals [14], and so on. Data handling is most crucial for IoT-based smart city. Based on the data analysis, expert system can make decisions in present circumstances. Data hierarchy at various layers in a smart city should maintain the privacy policies to sustain the trust among the entities of the environment. M. Sarkar et al. [15] in their research maintain the trust of a context-sensitive relational database. The remaining part of this chapter is as follows: Section 1.2 describes some previous relative works. Section 1.3 describes the overview of IoT with the basic architecture of the IoT communication network. Section 1.4 explains the application of IoT at various commercial fields. Proposed model is introduced in Section 1.5 followed by mathematical treatment in Section 1.6. Result and discussion are described in Section 1.7. Section 1.8 presents conclusion and future research scope.

1.2   RELATIVE WORK

IoT can be used by consumers in various fields, and some examples are smart cities, smart home, and elder care. These include connected vehicles, automation in homes, wearable device technology, healthcare, and appliances consisting remote monitoring [16]. IoT can be used for automation in cities, which can be turned into smart cities, and also home automation such as automatic lighting in cities and home, automatic air conditioning in summer season, automatic heating in winter season, enhanced security systems and antitheft system, automatic media access in smart home, and so on [17,18]. Sarkar et al. [19] proposed a trusted cloud service model for smart city using intelligence technique.

Some popular IoT devices for home applications are Bitdefender BOX IoT security solution, Google Home voice controller, Amazon Echo voice controller (second generation), Nest Cam Indoor camera, Mr. Coffee Smart Coffeemaker, SmartMat Intelligent Yoga Mat, Philips Hue, TrackR Bravo device, Linquet Bluetooth tracking sensor, Amazon Echo Spot smart alarm clock, BB8 SE Droid with force band, Nest smart thermostat, Amazon Echo Plus voice controller, Logitech Pop smart button controller, Nest Cam outdoor camera, AWS IoT Button programmable dash button, smart air quality monitor, smart household appliances, and so on. Some popular IoT devices for these applications are smart thermometer, blood pressure monitor, heart rate monitoring device, and so on [20]. Raafat et al. [21] in their research developed a smart home automation system. IoT can be used for taking care of elder persons at home with disabilities in health. IoT voice-controlled devices can be used to help elder persons at home. Sandeep et al. [22] described personalized heath recommendation system for people. Dudhe et al. explained the overview of IoT and various applications such as smart home, agriculture, healthcare, transportation, and so on. Also they mentioned some challenges for security issues in IoT [23]. Baker et al. surveyed about wearable healthcare system, challenges, and the future opportunities. Also they proposed a model for IoT healthcare monitoring system. Working of sensors such as photoplethysmographic pulse sensors, pressure-based pulse sensors, respiratory rate sensors, body temperature sensors, pulse oximetry sensors, and so on are explained. In addition, communication standards such as short-range communications and long-range communications standards are expanded [24]. Krishna and Nalini Sampath proposed a system to monitor patient health condition on real-time basis. Patients' health parameters such as body temperature, heart rate, percentage of oxygen saturation, and so on are fetched and transferred to the cloud and analyzed by authorized person using smartphone or laptop [25]. Chatterjee et al. proposed

IoT-based intelligent healthcare system to detect cardiovascular diseases. When a patient has cardiovascular diseases having abnormal heart rhythms, early detection of this can save the patient's life [26]. Fran et al. [27] in their article described how the concept of recommendation system could be applied within a smart city in the context of health application. In their paper, they considered collaborative improvement of the lifestyle quality.

1.3 OVERVIEW OF IOT

IoT platform consists of physical and information world. Physical world consists of device, gateway, communication channel (via gateway, without gateway, and direct communication), and so on, and information world indicates data that are accumulated from sensor or some different means. The collected data from the sensor are actuated, captured, stored, and processed by the system and then can be accessed by users. Data analysis is also done in different stages.

1.3.1 ARCHITECTURE OF IoT IN THE CONTEXT OF COMMUNICATION NETWORK

IoT reference model consists of four layers (Figure 1.1).
    The four layers are as follows:

- Application layer
- Service support and application support layer
- Network layer
- Device layer

1.3.1.1 Application Layer

The application layer contains IoT applications program.

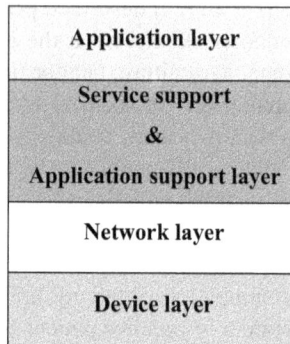

FIGURE 1.1   IoT reference model.

1.3.1.2 Service Support and Application Support Layer

Generic support: The generic support capabilities can be used by different IoT applications such as data processing/data storage.

Specific support: The specific support capabilities can be used for diversified applications.

1.3.1.3 Network Layer

Networking capabilities: Networking capabilities provide network connectivity, such as access control, resource control for transport, management of mobility, authentication, accounting, and authorization.

Transport capabilities: Transport capabilities concentrate on giving connectivity for transport of IoT service, data information, application, control, and management information.

1.3.1.4 Device Layer

Device capabilities: Direct interaction with the communication network: Without using gateway, devices are able to gather, receive, and upload information. Indirect interaction with the communication network: With the help of gateway, devices are able to gather, receive, and upload information.

Gateway capabilities: A gateway provides a place to preprocess that data locally at the edge before sending the data onto the cloud. An IoT gateway is a physical device or software program that serves as the connection point between the cloud and controllers, sensors, and intelligent devices.

1.4 APPLICATION OF IOT IN THE CONTEXT OF COMMERCIAL VIEW

IoT can possibly tame the weight of urbanization, make new understanding for city inhabitants, and make everyday living increasingly agreeable and secure.

1.4.1 MEDICAL AND HEALTHCARE

The Internet of Medical Things (IoMT) (likewise called the Internet of health things), is the utilization of the IoT for therapeutic and well-being-related purposes, information assortment and investigation for research, and checking [28–31]. IoT healthcare devices can be monitored through remote location using Internet. Some IoT medical and healthcare devices are fitbit wristband, hearing aids, heartbeat and rate monitoring device, blood pressure monitoring device, and so on.

1.4.2 TRANSPORTATION

IoT can be used in intelligent transportation system, smart GPS tracking, vehicle control, smart vehicle management, smart parking, smart traffic control, and smart electronic toll collection and road assistance [32–34].

Recently, the National Payments Corporation of India (NPCI) has developed the National Electronic Toll Collection (NETC) system to fulfill electronic tolling needs in India. They developed FASTag as payment mode at toll gates all over the India. FASTag is a device that uses RFID (radio frequency identification) technology for payments from moving vehicles. FASTag is affixed on the windscreen of the vehicle and enables to make the toll payments directly from the account which is linked to FASTag. This technology saves fuel and waiting time at the toll gate, and also we can monitor the account status and the deductions using mobile phone sitting at a remote location [35].

1.4.3 V2X Communications

In vehicular correspondence frameworks, vehicle-to-everything (V2X) correspondence comprises of three fundamental parts: vehicle-to-vehicle correspondence, vehicle-to-foundation correspondence, and vehicle-to-person on foot interchanges. V2X correspondence is the initial step to self-sufficient driving and associated street framework. The National Highway Traffic Safety Administration (NHTSA), an agency of the U.S. federal government, estimates that this technology can reduce the read accidents and traffic congestion. V2X-type communication can be implemented with the help of WLAN and cellular technology [36].

1.4.4 Agriculture

There are various IoT applications in farming [37,38] such as gathering data about rainfall, temperature, wind speed, humidity, infestation caused by various reasons, and content of the soil. These data can be used to automate farming techniques, which can be a new revolution in agriculture [39]. Nowadays, development of agricultural drone is quite popular, which can help in farming and monitoring of different entities in the field [40]. In August 2018, Toyota Tsusho started an organization with Microsoft to make fish-cultivating devices utilizing the Microsoft Azure application suite for IoT technologies associated to water management [41].

1.4.5 Energy Management

As home appliances or industry appliances, several devices are in use for daily life, and those devices are consuming energy from a power line, so monitoring of power consumption is needed in periodic time span. With the help of IoT monitoring, analysis of power consumption in house or industry is possible, and also the readings can be taken from a remote location. Several methods are proposed nowadays [42,43] for implementing energy management through IoT. This can also save energy consumption.

1.4.6 Living Lab

Living Lab combines and integrates research and innovation ideas with the help of public–private–people partnership. At present, several IoT-based Living Labs are providing innovative and technological products to the society. The IoT-based Living

Lab provides public networks, open infrastructure, and data for developers to accelerate emerging IoT innovations for smart city solutions [44,45].

1.4.7 INTERNET OF BATTLEFIELD THINGS

The Internet of Battlefield Things (IoBT) is a venture begun by the U.S. ARL (Army Research Laboratory) that spotlights on the essential science identified with IoT that lift the capacities of fighters [46]. In 2017, ARL propelled the IoBT Collaborative Research Alliance (IoBT-CRA) and made a working cooperation among army specialists, industry, and college to propel advances in army activities [47–50].

1.4.8 OCEAN OF THINGS

The ocean of things venture is a DARPA (Defense Advanced Research Projects Agency)-driven program intended to build up an IoT crosswise over huge sea regions for the motivations behind gathering, checking, and investigating ecological and vessel action information. The venture involves the sending of around 50,000 buoys that house a passive sensor suite that self-sufficiently identifies and tracks military and business vessels as a major aspect of a cloud-based network [51].

1.5 PROPOSED MODEL

Smart and efficient medical diagnostic system is proposed for smart city based on IOT. Due to this rapid growing world, people are not able to concentrate on their physical fitness by visiting physicians at regular intervals. Present era is fully controlled by Internet at every field of work. The concept of IOT-based application is proposed for identifying the diseases.

Let us discuss smart social healthcare and the key innovations. Smart health technology consolidates smart technology and the most recent cell phones with well-being features. These days, various activities have been intended to energize a more extensive perspective on well-being and prosperity; therefore, wearable gadgets such as wellness tracker or wellness groups and even well-being evaluation applications in cell phones have increased attraction among wellness lovers. These gadgets are keen in the sense as they simply screen well-being and give solutions if necessary at the correct time. Brilliant gadgets go about as the base of savvy medicinal services. Smart health innovation communicates and draws information delivered by those gadgets, which can be broken down by specialists, analysts, and medicinal services experts for better customized decisions and arrangements. These advanced records reduce expense and time of patients and clinics as they offer customized medicines and meds as well as give preventive measures through real-time data collection. A mobile network architecture/mobile environment is shown in Figure 1.2.

The mobile network architecture consists of various components such as mobile phone, mobile switching center (MSC), base transceiver (BT), base station

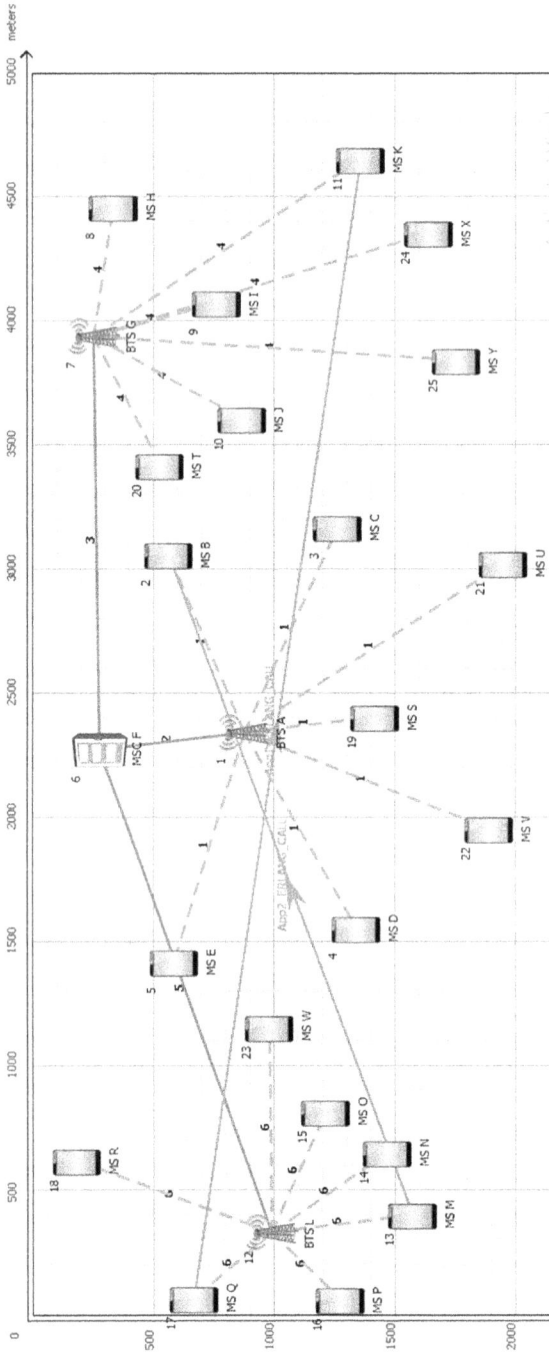

FIGURE 1.2   Mobile environmental in IoT.

controller (BSC), gateway, etc. This architecture provides PSTN and Internet services to the user. If data transfer takes place between 17 and 11 after simulation in NetSim software, we can see the network performance and the throughput of the network. The throughput can be increased by selection of proper data transfer protocol. A real health data is used to validate the model. At the beginning, through a statistical approach, the disease and nondisease percentage of total data set is determined. Fuzzy reasoning is used to arrive at a smart decision.

1.6 MATHEMATICAL TREATMENT

To achieve the proposed model, two types of approaches are deployed. In this chapter, statistical approach and fuzzy reasoning methodologies are deployed. Statistical approach is used to determine the percentage of persons with disease and no disease. Optimistic decision is achieved by using fuzzy reasoning methodology.

1.6.1 STATISTICAL APPROACH

Let us consider total sample data space S, represented by ith number of disjoint subsets Z_i, such that $S = U_i Z_i$ with $Z_i \cap Z_j = \varphi$ for $i \neq j$. Also consider $P(Z_i) \neq 0$ $\forall i$. Any arbitrary subset G can be denoted as $G = G \cap S = G \cap (U_i Z_i) = U_i (G \cap Z_i)$ where $G \cap Z_i$ is disjoint, and their probability as equation (1.1):

$$p(G) = P(U_i(G \cap Z_i)) = \sum_i P(G \cap Z_i) = \sum_i P(G \mid Z_i) P(Z_i) \qquad (1.1)$$

Now, sample space is split into Z_i subsets. The probability of Z_i is evaluated by Bayes' theorem as equation (1.2):

$$P(Z \mid G) = \frac{P(G \mid Z) P(Z)}{\sum_i P(G \mid Z_i) P(Z_i)} \qquad (1.2)$$

As per the prior knowledge of probability of the disease which carried out the percentage of C which is the total population size. If a test yields a positive result of probability F%, then

$$P(+ \mid disease) = F / 100$$

$$P(- \mid disease) = C / 100 \text{ where, } F \text{ and } C \text{ are both positive numbers} \qquad (1.3)$$

According to the Bayes' theorem, equation (1.2) is rewritten as equation (1.4)

$$P(disease \mid +) = \frac{P(+ \mid disease) P(disease)}{P(+ \mid disease) P(disease) + P(+ \mid no disease) P(no disease)} \qquad (1.4)$$

1.6.2   FUZZY REASONING

Fuzzy reasoning is deployed to overcome the uncertainty of no disease state of the total sample space.

Consider M and N are both fuzzy linguistic values defined by fuzzy set X and Y, respectively.

If x is M, then y is N. Now A is coupled with B, then,

$$R = M \Rightarrow N = M * N = \int_{X*Y} \mu_M(x) * \mu_N(y) / (x, y) \qquad (1.5)$$

where $*$ is a T-normoperator.

As per minimum operator proposed by Mamdani [52]

$$R_{dp} = M * N = \int_{X*Y} \mu_M(x) * \mu_N(y) / (x, y) \qquad (1.6)$$

where $f(m,n) = m^\wedge n = \begin{cases} m & \text{if } n = 1 \\ n & \text{if } m = 1 \\ 0 & \text{otherwise} \end{cases}$

Conclusion: If Z is C, then $M * N \rightarrow C$

$$R(M, N, C) = (M * N) * C = \int_{X*Y*Z} \mu_M(x) \wedge \mu_N(y) \wedge \mu_C(Z) / (x, y, z) \qquad (1.7)$$

Fuzzy projection method is applied for implementing expert decision-making.

Let Y be a fixed set and M be a fuzzy set in Y. Fuzzy set is an object of Y in the form as equation (1.8):

$$M = \left\{ (y, \mu_M(y), v_M(y)) \middle| y \in Y \right\} \qquad (1.8)$$

where $\mu_M(y) : Y \rightarrow [0,1]$ and $v_M(y) : Y \rightarrow [0,1]$

Degree of membership value and degree of nonmembership value are expressed by equation (1.8).

Assume $Y = \{y_1, y_2, \ldots, y_m\}$ be a finite set in universe of discourse. Set M and N are two IFs in Y; then projection of M on N can be expressed as equation (1.9):

$$(M \downarrow N) = \frac{1}{|N|} \sum_{i=1}^{n} \left(\mu_{\alpha_i} . \mu_{\beta_i} + v_{\alpha_i} . v_{\beta_i} + \pi_{\alpha_i} . \pi_{\beta_i} \right) \qquad (1.9)$$

where $\alpha_i = \mu_{\alpha_i}.v_{\alpha_i}.\pi_{\alpha_i}$ and $\beta_i = \mu_{\beta_i}.v_{\beta_i}.\pi_{\beta_i}$. Equation (1.9) expresses the ith IFNs of M and N, respectively.

More than one rule is defined in the proposed model. Therefore, individual weight is applied to each rule. After applying weight function, projection of M on N is defined as equation (1.10).

$$(M \downarrow N) = \frac{1}{|N|_w} \sum_{i=1}^{n} w_i^2 \left(\mu_{\alpha_i}.\mu_{\beta_i} + v_{\alpha_i}.v_{\beta_i} + \pi_{\alpha_i}.\pi_{\beta_i} \right) \qquad (1.10)$$

Vector $W = \{w_1, w_2, w_3, \ldots, w_n\}$ of $y_j \forall j = 1,2,3,\ldots,n$ and $W_j \notin [0,1]$, $\sum_{j=1}^{n} W_j = 1$.

A collaborative decision-making is considered in this proposed model. A set of decision-making $C = \{c_1, c_2, c_3, \ldots, c_n\}$. Let us consider p number of decisions are considered. The collaborative decision can be determined as equation (1.11):

$$C_{ij}^{(l)} = \left\{ \mu_{ij}^{(l)}, v_{ij}^{(l)}, \pi_{ij}^{(l)} \right\} \qquad (1.11)$$

Mean of the decision is evaluated by l number of decisions as equation (1.12):

$$C_{ij}^{*} = \left\{ \mu_{ij}^{*}, v_{ij}^{*}, \pi_{ij}^{*} \right\} \qquad (1.12)$$

where μ_{ij}^{*}, v_{ij}^{*}, and π_{ij}^{*} are defined as $\mu_{ij}^{*} = \frac{1}{l} \sum_{p=1}^{l} \mu_{ij}^{(p)}$, $v_{ij}^{*} = \frac{1}{l} \sum_{p=1}^{l} v_{ij}^{(p)}$, and $\pi_{ij}^{*} = \frac{1}{l} \sum_{p=1}^{l} \pi_{ij}^{(p)}$.

The weight

$$W_{ij}^{(p)} = \frac{[C_{ij}^{(p)} \downarrow C_{ij}^{*}]}{\sum_{p=1}^{l} [C_{ij}^{(p)} \downarrow C_{ij}^{*}]} \qquad (1.13)$$

where $1 \le i \ge m$, $1 \le j \ge n$, and $1 \le p \ge l \ \forall \ m \ne n \ne p$.

Aggregation operation is deployed to find the collective decision matrix D.

$$D_{ij} = W_{ij}^{(1)} C_{ij}^{(1)} + W_{ij}^{(2)} C_{ij}^{(2)} + \cdots + W_{ij}^{(l)} C_{ij}^{(l)} \qquad (1.14)$$

$$D = (D_{ij})_{m \times n} \qquad (1.15)$$

Fuzzy rules are developed based on training set $S : \{X_{ij}; Y_{ij}\}$ where $i = 1, 2, \dots, m$ and $j = 1, 2, \dots, n$.

Every data point is in X_{ij}. The maximum value of D_{ij} is calculated by equation (1.16):

$$D_{\max} = \max_{i=1,2,\dots,n} \left(\mu_M(x_{ij})\right) \tag{1.16}$$

1.6.3   FUZZY RULES SET

In this chapter, six various diseases and their corresponding symptoms are considered to validate the proposed model: fever (FV), fungal infection (FI), allergy (AL), drug reaction (DR), diabetes (DI), and jaundice (JA). Twenty two different rules are considered for these six diseases, and corresponding suggestions for all these rules are also defined.

Fever

> **Rule 1: IF** FV $\leq$ 95°F && days $\leq$ 2, **THEN** normal fever condition.
> **Suggestion** – Then suggest to take paracetamol.
> **Rule 2: IF** FV $\geq$ 100°F && 4 days $\leq$ days $\geq$ 6 days, **THEN** not normal fever.
> **Suggestion** – Take paracetamol and antibiotic pills.
> **Rule 3: IF** FV $\geq$ 100°F && days $\geq$ 6, **THEN** not normal fever.
> **Suggestion** – Then suggest blood test for safety purpose and paracetamol + antibiotic pills.
> **Rule 4: IF** FV $\geq$ 102°F && days $\geq$ 6, **THEN** high fever.
> **Suggestion** – Then visit the nearby hospital/suggested hospital.

Fungal Infection

> **Rule 1: IF FI** = itching+ skin rashes && days $\geq$ 2, **THEN** normal infection.
> **Suggestion** – Then suggest some clotrimazole or econazole.
> **Rule 2: IF FI** = itching + skin rash + nodal skin && days $\geq$ 4, **THEN** medium infection.
> **Suggestion** – Then suggest some fluconazole and amphotericin.
> **Rule 3: IF FI** = itching + skin rashes + nodal skin && days $\geq$ 6, **THEN** high infection.
> **Suggestion** – Then suggest some ketoconazole + antifungal antibiotic.
> **Rule 4: IF FI** = itching + nodal skin + fever && days $\geq$ 8, **THEN** high infection.
> **Suggestion** – Then suggest to visit nearby hospital.

Allergy
Common for all – Check the food habits

> **Rule 1**: IF AL days $\geq$ 3, **THEN** normal allergy
> **Suggestion** – Then suggest alive and dixon + antibiotic.
> **Rule 2:** IF AL days $\geq$ 7, **THEN** medium allergy
> **Suggestion** – Then suggest blood test and appropriate antibiotics + anti histamine

Rule 3: IF AL days ≥ 16, **THEN** high allergy
Suggestion – Suggest to visit hospital

Drug Reaction

Rule 1: IF DR a fast-spreading painful red or blistered area on the skin &&
days ≥ ww5, **THEN** primary stage.
Suggestion – Then suggest to give an antihistamine + use cool compress on the
area or have the person take cool shower + avoid strong soaps, detergents,
and other chemicals.
Rule 2: IF DR discomfort && fever && days ≥ 10, **THEN** primary stage
Suggestion – Stay in a cool room + wear loose fitting clothes + apply calamine
on the rashes area.
Rule 3: IF DR increased heat rate && heightened blood pressure, **THEN**
medium stage.
Suggestion – Take diuretics + beta-blockers.
Rule 4: IF DR very high body temperature + muscle shakes or tremors +
agitation, **THEN** highly addicted.
Suggestion – Then consult to a doctor immediately.

Diabetes

Rule 1: IF DI urinating often && feeling very thirsty &&days ≤ 2, **THEN**
initial stage of diabetes.
Suggestion – Have more water + wait 4/5 days.
Rules 2: IF DI feeling hungry, even though you are eating && extreme
fatigue && days ≤ 4, **THEN** initial stage.
Suggestion – Keep in observation for some days.
Rules 3: IF DI blurry vision && weight loss && tingling, **THEN** severe diabetes.
Suggestion – Then go to blood test.
Rules 4: IF DI above the range, **THEN** severe diabetes.
Suggestion – Then consult to a doctor.

Jaundice

Rule 1: IF JA fatigue && abdominal pain, **THEN** primary stage.
Suggestion – Then take some pain killer + wait some days.
Rule 2: IF JA weight loss && vomiting, **THEN** medium stage.
Suggestion – Take some phentermine + topiramate or lorcaserin.
Rule 3: IF JA body turns yellow && eyes are yellow, **THEN** severe stage.
Suggestion – Visit to nearest hospital and consult to doctor.

1.6.4 Proposed Algorithm

In this research work, two algorithms are proposed. Algorithm 1.1 is designed
based on statistical Bayes' theorem to determine the disease and nondisease
percentage.

**Algorithm 1.1 Statistical Approach to Determine
the Disease and No-Disease State**

Input: Real-time data set as a sample space S.
Output: Find out with disease and with no-disease condition.
Step 1: Begin.
Step 2: Initialization of the sample space S.
Step 3: Split the sample space with i number of disjoint subsets.
Step 4: Each subset Z_i such that $P(Z_i) \neq 0 \ \forall i$.
Step 5: Deploying Bayes' theorem to determine probability of Z_i:

$$P(Z \mid G) = \frac{P(G \mid Z) P(Z)}{\sum\limits_i P(G \mid Z_i) P(Z_i)}$$

Step 6: Calculating the percentage of disease and no-disease state as equation (1.4):

$$P(\text{disease} \mid +) = \frac{P(+ \mid \text{disease}) P(\text{disease})}{P(+ \mid \text{disease}) P(\text{disease}) + P(+ \mid \text{no disease}) P(\text{no disease})}$$

Step 7: End.

Algorithm 1.2 is designed based on fuzzy reasoning. To determine the optimistic decision from collaborative decision, a maximum optimistic decision is determined.

Algorithm 1.2 Determine the Optimistic Collective Decision

Input: Sample data set as a finite fuzzy sets with linguistic values.
Output: Optimistic collective decision.
Step 1: Begin.
Step 2: Initialization: the M and N are both fuzzy linguistic values defined by
fuzzy set X and Y, respectively
Step 3: Deploying fuzzy Mamdani as equation (1.6):

$$R_{dp} = M * N = \int_{X*Y} \mu_M(x) * \mu_N(y) / (x, y)$$

Step 4: Deploying fuzzy projection method to determine the expert decision
as equation (1.8):

$$M = \left\{ \left(y, \mu_M(y), v_M(y) \right) \middle| y \in Y \right\}$$

Step 5: Applying fuzzy IF's rules, then projection of M on N can be expressed
as equation (1.9):

$$(M \downarrow N) = \frac{1}{|N|} \sum_{i=1}^{n} \left(\mu_{\alpha_i} . \mu_{\beta_i} + v_{\alpha_i} . v_{\beta_i} + \pi_{\alpha_i} . \pi_{\beta_i} \right)$$

Step 6: Weight vector W is assigned: $W_{ij}^{(p)} = \dfrac{[C_{ij}^{(p)} \downarrow C_{ij}^{*}]}{\sum\limits_{p=1}^{l} [C_{ij}^{(p)} \downarrow C_{ij}^{*}]}$

Step 7: The collaborative decision can be determined as $C_{ij}^{(l)} = \left\{ \mu_{ij}^{(l)}, v_{ij}^{(l)}, \pi_{ij}^{(l)} \right\}$.

Step 8: Mean of the decision is evaluated by l number of decision as equation (1.12):

$$C_{ij}^{*} = \left\{ \mu_{ij}^{*}, v_{ij}^{*}, \pi_{ij}^{*} \right\}$$

Step 9: Aggregation operation is deployed to find the collective decision matrix D:

$$D_{ij} = W_{ij}^{(1)} C_{ij}^{(1)} + W_{ij}^{(2)} C_{ij}^{(2)} + \cdots + W_{ij}^{(l)} C_{ij}^{(l)}$$

$$D = (D_{ij})_{m \times n}$$

The maximum value of D_{ij} $D_{max} = \max\limits_{i=1,2,\ldots,n} \left(\mu_M(x_{ij}) \right)$.

Step 10: End.

1.6.5 DATA PREPARATION

A real data set for disease prediction using machine learning [16] is used to validate the proposed model. In this research, 15 patients are considered with their 10 various symptoms of disease used as a test bed. Table 1.1 describes the patients and their various symptoms.

To validate the proposed model, various threshold values are fixed for various diseases, and also their corresponding fuzzy membership values are fixed (Table 1.2).

TABLE 1.1

Patients and Their Various Symptoms

Patient	Skin Rash	Itching	Shivering	Chills	Fatigue	Diarrhea	Mild fever	Yellowing eyes	Prognosis
1	1	1	0	0	0	0	0	0	Fungal infection
2	0	0	1	1	0	0	0	0	Allergy
3	1	1	0	0	0	0	0	0	Drug reaction
4	1	1	0	0	0	0	0	0	Drug reaction
5	0	1	0	0	1	0	0	0	Jaundice
6	0	0	0	1	0	1	0	0	Malaria
7	1	1	0	0	1	0	1	0	Chicken pox
8	1	0	0	1	1	0	0	0	Dengue
9	0	0	0	1	1	0	0	1	Typhoid
10	0	1	0	0	1	0	0	1	Hepatitis B

TABLE 1.2

Threshold Value of Various Diseases and Corresponding Fuzzy Membership Value

Sl No	Name of the Diseases	Threshold Value	Fuzzy Membership Value
1	Fever	$\geq 95°F$	1
2	Fungal infection	Normal (N), medium (M), high (H)	$N = 0, M = 0.5, H = 0.98$
3	Allergy	Normal (N), medium (M), high (H)	$N = 0.12, M = 0.55, H = 0.87$
4	Drug reaction	Primary (P), medium (M), high (H)	$P = 0.1, M = 0.65, H = 0.92$
5	Diabetes	Initial stage (I), severe (S)	$I = 0.2, S = 0.96$
6	Jaundice	Primary stage (P), medium (M), severe (S)	$P = 0.13, M = 0.4, S = 0.94$

1.7 RESULTS AND DISCUSSION

NetSim software is used to simulate the mobile-based IoT environment, which is shown in Figure 1.2. The simulation results are shown in Figures 1.3–1.5.

The transmission link throughput graph is shown in Figure 1.3. The throughput is increasing with respect to time, and after certain time, almost constant values are observed in the graph.

The Erlang call throughput graph is shown in Figure 1.4. It is observed that the throughput is increasing exponentially with respect to time because continuous calls are arrived.

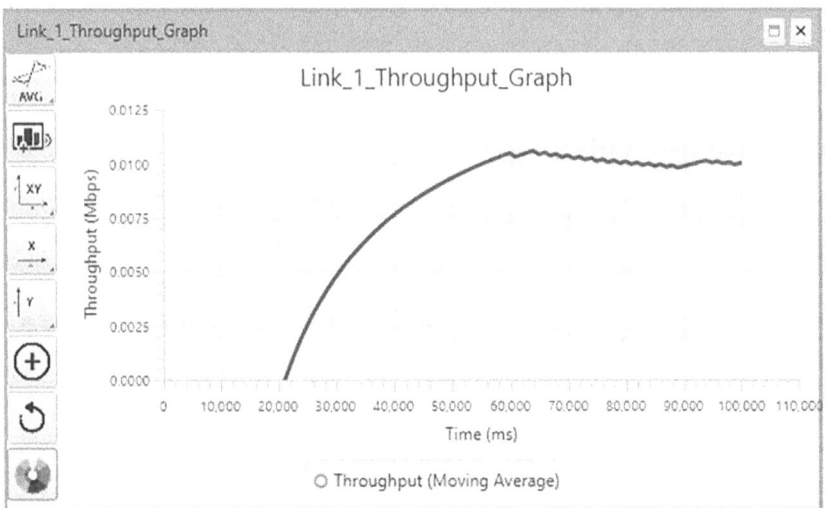

FIGURE 1.3 Link 1 throughput graph.

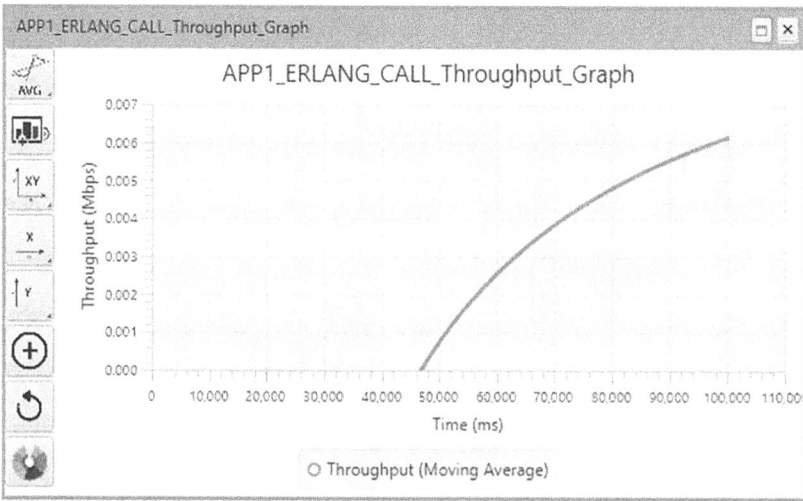

FIGURE 1.4 Erlang call throughput graph.

Application Id	Throughput Plot ▲	Application Name	Packet transmitted	Packet received	Throughput (Mbps)	Delay(microsec)
1	Application throughput plot	APP1_ERLANG_CALL	702	480	0.006144	18063764.283380
2	Application throughput plot	APP2_ERLANG_CALL	3002	712	0.009114	32400325.521587

Application_metrics   ☐ Detailed View

Application_Metrics_Table

FIGURE 1.5 Application metrics.

From the application metrics table, it is observed that the total number of packet transmitted, the total number of packet received, throughput, and the delay take place in the network.

To validate the proposed model, a real data set [53] is used. The data set is randomly split into two parts. From the data set, 40% of randomly selected data are used to train the model, and remaining 60% of data are used for testing. MatLab R2018a is used for result and simulation.

Figure 1.6 describes the membership grade value of various diseases for different patients. Figure 1.6 segregates different diseases based on their symptoms at the data set. The data set consists of only the value of various symptoms of different patients. After analyzing the data set, total percentage of people suffering from disease or not are determined.

Figure 1.7 is presented as a spiral form. In this figure, the rate of suffered person is shown. Different colors indicate the different disease. Figure 1.7 is plotted based on the training data set.

Based on testing data set, Figure 1.8 is plotted. Sixty percent of data are used for testing purpose. Figure 1.8 shows different people suffering from similar disease, and their symptoms are similar. To determine the optimistic decision of the disease,

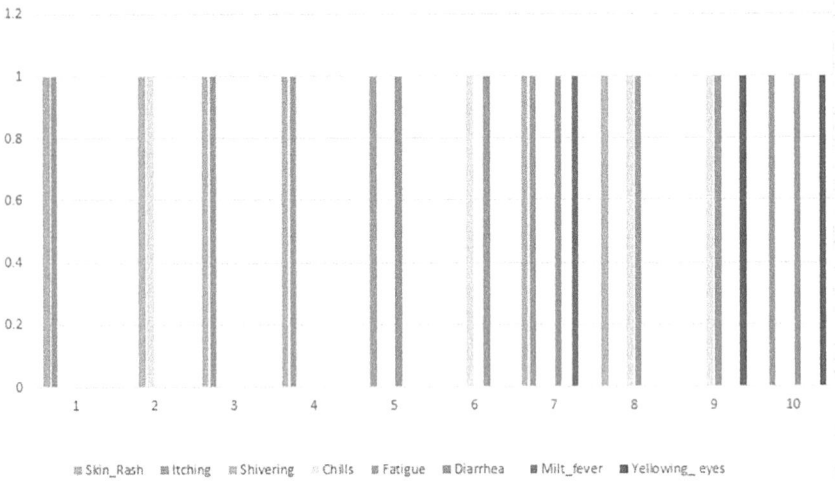

FIGURE 1.6   Various diseases as per different patients with membership grade.

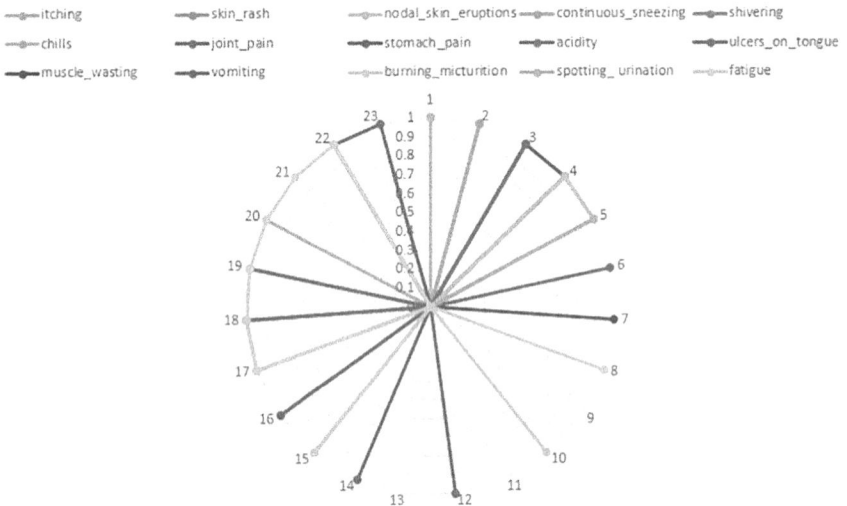

FIGURE 1.7   Various symptoms with their threshold value and the number of patients affected.

expert system is used. The simulation result for optimistic decision for the disease is shown in Figure 1.9. Due to slightly changes of the symptom, a patient is considered for suffering different disease by the medical practitioner.

In this chapter, the maximum value from the collaborative decision is determined. Figure 1.4 describes the maximum value from the collaborative maximize values. After successful testing of the proposed model, the symptoms of ten individual patients are tested. Among the ten patients, six patients are suffering from various

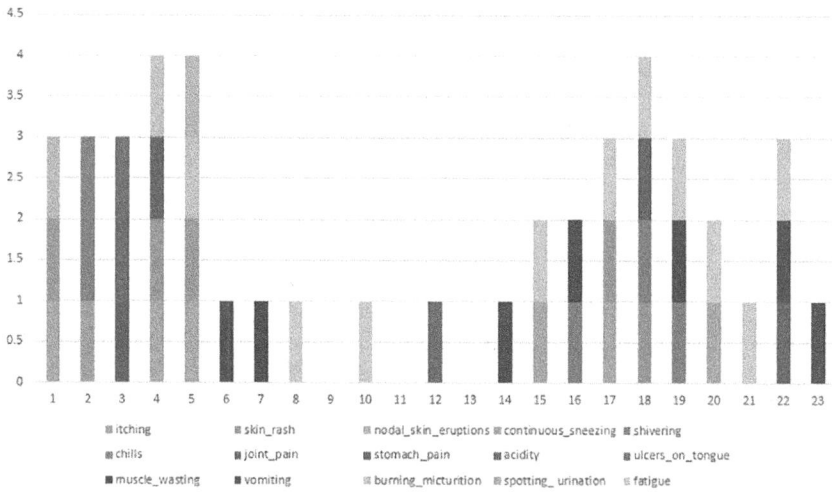

FIGURE 1.8   Various symptoms and their corresponding patients.

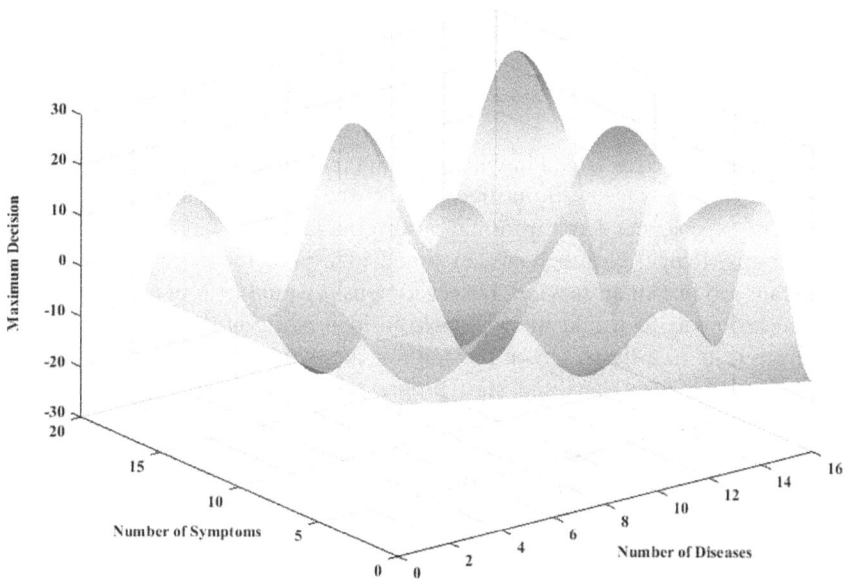

FIGURE 1.9   Maximum decision value of collaborative optimistic decision.

diseases. Patients and their corresponding disease, diagnosed by the proposed model, are shown in Table 1.3.

To measure the efficiency, precision and recall are evaluated and shown in Table 1.4.

In this research experiment, the proposed model is tested by three iterations. As per Table 1.4, the overall precision and recall are 89.94% and 88.68%, respectively.

TABLE 1.3

Tested Results from the Proposed Model

Sl No	Name of the User	Disease
1	User 1	Fungal infection but in medium stage
2	User 2	Allergy problem
3	User 3	Highly addicted
4	User 4	Severe diabetes
5	User 5	Fungal infection but highly infected
6	User 6	High fever

TABLE 1.4

Performance Analysis of the Proposed Model

Training and Testing Strategy	Strategy 1 (60%–40%)	Strategy 2 (70%–30%)	Strategy 3 (80%–20%)
Precision	89.34%	88.02%	92.46%
Recall	88.16%	86.57%	91.32%

1.8   CONCLUSION

IoT in medical services is tied in with releasing the intensity of associated gadgets and sensors that are generally utilized in the division. IoT can be utilized to get significant experiences from information originating from fetal screens, electrocardiograms, temperature screens, or blood glucose screens. IoT can assume a significant job in human services observation as it would help in early location of medical problems. It would likewise help in incorporating the information gathered from tests in a flash, screen the state of the patient, and afterward hand off that data to the specialists and staff progressively, consequently improving the effectiveness in the general human services framework. In this chapter, an expert system is proposed to determine a maximum optimistic decision from a collaborative decision. The proposed expert system is deployed for medical decision-making in a smart city. Statistical approach and fuzzy reasoning methodologies are used to achieve the goal. Deep learning technique would be used to enhance the precision and recall value in the context of big data, which will be considered as future research work.

REFERENCES

1. https://en.wikipedia.org/wiki/Internet_of_things. Retrieved January 4, 2020.
2. Rouse, Margaret (2019). "Internet of Things (IoT)." IOT Agenda. Retrieved January 4, 2020.

3. Brown, Eric (September 20, 2016). "21 Open Source Projects for IoT." Linux.com. Retrieved January 4, 2020.

4. ITU. "Internet of Things Global Standards Initiative." Retrieved January 4, 2020.

5. Hendricks, Drew. "The Trouble with the Internet of Things." London Data store. Greater London Authority. Retrieved January 4, 2020.

6. ITU. "Overview of the Internet of Things." Recommendation ITU-T Y.2060, June 2012.

7. Haase, Jan, Alahmad, Mahmoud, Nishi, Hiroaki, Ploennigs, Joern, Tsang, Kim Fung, "The IOT mediated built environment: A brief survey," *2016 IEEE 14th International Conference on Industrial Informatics*, pp. 1065–1068, 2016. doi:10.1109/INDIN.2016.7819322.

8. Karlgren, Jussi, Fahlén, Lennart, Wallberg, Anders, Hansson, Pär, Ståhl, Olov, Söderberg, Jonas, Åkesson, Karl-Petter, *Socially Intelligent Interfaces for Increased Energy Awareness in the Home.* The Internet of Things. Lecture Notes in Computer Science. 4952. Springer. pp. 263–275, 2008. doi:10.1007/978-3-540–78731-0_17.

9. Rico, Juan (April 22–24, 2014). "Going beyond monitoring and actuating in large scale smart cities." NFC & Proximity Solutions – WIMA Monaco.

10. Kodali, Ravi Kishore, Mandal, Snehashish, "IoT based weather station," *International Conference on Control, Instrumentation, Communication and Computational Technologies (ICCICCT)*, 2016.

11. Use case: Sensitive wildlife monitoring. https://web.archive.org/web/20140714124750/http://fit-equipex.fr/use-cases/23-use-case-sensitive-wildlife-monitoring. Retrieved January 4, 2020.

12. Liu, Xiaohan, Yang, Tao, Yan, Baoping, "Internet of Things for wildlife monitoring," *IEEE/CIC International Conference on Communications in China - Workshops (CIC/ICCC)*, 2015.

13. Nóbrega, Luís, Tavares, André, Cardoso, António, Gonçalves, Pedro, "Animal monitoring based on IoT technologies," *IoT Vertical and Topical Summit on Agriculture - Tuscany (IOT Tuscany)*, 2018.

14. Giordano, Stefano, Seitanidis, Ilias, Ojo, Mike, Adami, Davide, Vignoli, Fabio, "IoT solutions for crop protection against wild animal attacks," *IEEE International Conference on Environmental Engineering (EE)*, 2018.

15. Sarkar, Manash, Banerjee, Soumya, Hassanien, Aboul Ella, "Evaluating the Degree of Trust under Context Sensitive Relational Database Hierarchy Using Hybrid Intelligent Approach," *International Journal of Rough Sets and Data Analysis (IJRSDA)*, 2 (1), 1–21, 2015.

16. "How IoTs Are Changing the Fundamentals of 'Retailing'." Trak.in – Indian Business of Tech, Mobile & Startups. August 30, 2016. Retrieved January 4, 2020.

17. Domb, Menachem (February 28, 2019). Smart Home Systems Based on Internet of Things, Internet of Things (IoT) for Automated and Smart Applications, Yasser Ismail, IntechOpen, 2019.

18. https://www.intel.in/content/www/in/en/internet-of-things/smart-home.html. Retrieved January 4, 2020.

19. Sarkar, Manash, Banerjee, Soumya, Badr, Yoaukim, Sangaiah, Arun Kumar, "Configuring a Trusted Cloud Service Model for Smart City Exploration Using Hybrid Intelligence," *International Journal of Ambient Computing and Intelligence*, 8 (3), 1–21, 2017.

20. "Overview of the Most Popular Smart Home Devices." http://iotlineup.com. Retrieved January 4, 2020.

21. Aburukba, Raafat, Al-Ali, A. R., Kandil, Nourhan, Abu Damis, Diala (May 10, 2016). Configurable ZigBee-based control system for people with multiple disabilities in smart homes. pp. 1–5. doi:10.1109/ICCSII.2016.7462435.

22. Sandeep Kumar, M., Prabhu, J., "Healthcare recommender system based on Smart–health Routes," *International Conference for Phoenixes on Emerging Current Trends in Engineering and Management (PECTEAM 2018), Advances in Engineering Research (AER)*, volume 142, pp. 42–46, 2018.

23. Dudhe, P.V., Kadam, N.V., Hushangabade, R.M., Deshmukh, M.S., "Internet of Things (IOT): An overview and its applications," *International Conference on Energy, Communication, Data Analytics and Soft Computing*, 2017.

24. Baker, Stephanie B., Xiang, Wei, Atkinson, Ian, "Internet of Things for Smart Healthcare: Technologies, Challenges, and Opportunities," *IEEE Access*, 5, 26521–26544, 2017.

25. Krishna, C. S., Sampath, Nalini, "Healthcare monitoring system based on IoT," *International Conference on Computational Systems and Information Technology for Sustainable Solution*, 2017.

26. Chatterjee, Parag, Cymberknop, Leandro J., Armentano, Ricardo L., "IoT-based decision support system for intelligent healthcare applied to cardiovascular diseases," *International Conference on Communication Systems and Network Technologies*, 2017.

27. Casino, Fran, Patsakis, Constantinos, Batista, Edgar, Postolache, Octavian, Martínez-Ballesté, Antoni, Solanas, Agusti, "Smart Healthcare in the IoT Era: A Context-Aware Recommendation Example," *International Symposium in Sensing and Instrumentation in IoT Era (ISSI)*, Shanghai, China, September 6–7, 2018.

28. da Costa, Cristiano Andréda, Pasluosta, Cristian F., Eskofier, Björn, da Silva, Denise Bandeira, da Rosa Righi, Rodrigo, "Internet of Health Things: Toward intelligent vital signs monitoring in hospital wards," *Artificial Intelligence in Medicine*, 89, 61–69, 2018. doi:10.1016/j.artmed.2018.05.005. PMID 29871778.

29. Engineer, Altaf, Sternberg, Esther M., Najafi, Bijan, "Designing Interiors to Mitigate Physical and Cognitive Deficits Related to Aging and to Promote Longevity in Older Adults: A Review," *Gerontology*, 64 (6), 612–622, 2018. doi:10.1159/000491488.

30. Gatouillat, Arthur, Badr, Youakim, Massot, Bertrand, Sejdic, Ervin, "Internet of Medical Things: A Review of Recent Contributions Dealing with Cyber-Physical Systems in Medicine," *IEEE Internet of Things Journal*, 5 (5), 3810–3822, 2018. doi:10.1109/jiot.2018.2849014. ISSN 2327–4662.

31. Topol, Eric, *The Patient Will See You Now: The Future of Medicine Is in Your Hands*. Basic Books, 2016,

32. Meneguette, Rodolfo I., De Grande, Robson E., Loureiro, Antonio A. F., *Intelligent Transport System in Smart Cities*. Springer Nature Customer Service Center LLC, 2018.

33. Helfert, Markus, Klein, Cornel, Donnellan, Brian, Gusikhin, Oleg, "Smart Cities, Green Technologies, and Intelligent Transport Systems," *5th International Conference, SMARTGREENS 2016, and Second International Conference*, VEHITS 2016, Rome, Italy.

34. "Key Applications of the Smart IoT to Transform Transportation." September 20, 2016, http://www.wiomax.com/what-can-the-smart-iot-transform-transportation-and-smart-cities. Retrieved January 4, 2020.

35. National Electronic Toll Collection, https://www.npci.org.in/netc. Retrieved January 4, 2020.

36. Vehicle-to-Everything, https://en.wikipedia.org/wiki/Vehicle-to-everything. Retrieved January 4, 2020.

37. Meola, A. (2016). "Why IoT, big data & smart farming are the future of agriculture." Business Insider. Insider, Inc. Retrieved January 4, 2020.

38. Muangprathuba, Jirapond, Boonnama, Nathaphon, Kajornkasirata, Siriwan, Lekbangponga, Narongsak, Wanichsombata, Apirat, Nillaorb, Pichetwut, "IoT and Agriculture Data Analysis for Smart Farm," *Computers and Electronics in Agriculture*, 156, 467–474, 2019.

39. "IoT Applications in Agriculture the demand for Growing Population Can Be Successfully Met with IoT." https://www.iotforall.com/iot-applications-in-agriculture/. Retrieved January 4, 2020.

40. Reinecke, Marthinus, Prinsloo, Tania, "The influence of drone monitoring on crop health and harvest size," *2017 1st International Conference on Next Generation Computing Applications (NextComp)*, 2017.

41. "Google Goes Bilingual, Facebook Fleshes Out Translation and TensorFlow Is Dope – and, Microsoft Is Assisting Fish Farmers in Japan." https://www.theregister. co.uk/2018/09/01/ai_roundup_310818/. Retrieved January 4, 2020.

42. Karthikeyan, S., Bhuvaneswari, P. T. V., "IoT based real-time residential energy meter monitoring system," *2017 Trends in Industrial Measurement and Automation (TIMA)*, 2017.

43. Prathik, M., Anitha, K., Anitha, V., "Smart energy meter surveillance using IoT," *International Conference on Power, Energy, Control and Transmission Systems (ICPECTS)*, 2018.

44. Wellsandt, Stefan, Kalverkamp, Matthias, Eschenbächer, Jens, Thoben, Klaus-Dieter, "Living Lab Approach to Create an Internet of Things Service," *International ICE Conference on Engineering, Technology and Innovation*, 2012.

45. Tsai-Lin, Tung-Fei, Chang, Yang-Yi, "Framing a smart service with living lab approach: A Case of introducing mobile service within 4G for smart tourism in Taiwan," *IEEE International Conference on Engineering, Technology and Innovation (ICE/ITMC)*, 2018.

46. "Army Takes on Wicked Problems with the Internet of Battlefield Things." MeriTalk. January 30, 2018. Retrieved January 4, 2020.

47. Gudeman, Kim (2017). "Next-Generation Internet of Battle Things (IoBT) Aims to Help Keep Troops and Civilians Safe." ECE Illinois. Retrieved January 4, 2020.

48. "Internet of Battlefield Things (IOBT)." CCDC Army Research Laboratory. Retrieved October 31, 2019.

49. Cameron, Lori, "Internet of Things Meets the Military and Battlefield: Connecting Gear and Biometric Wearables for an IoMT and IoBT," *IEEE Computer Society*, 2018.

50. Russell, Stephen, Abdelzaher, Tarek, "The Internet of Battlefield Things: The next generation of command, control, communications and intelligence (C3I) decision-making," *IEEE Military Communications Conference (MILCOM)*, 2018.

51. "DARPA Floats a Proposal for the Ocean of Things." MeriTalk. January 3, 2018. Retrieved January 4, 2020.

52. Ross, Timothy J., *Fuzzy Logic with Engineering Applications* (2nd ed.). John Wiley & Sons, 2005.

53. http://www.kaaggle.com/meelima98/disease-prediction-using-machine-learning. Accessed on October 10, 2019.

2 Soft Computing Techniques for Boolean Function and Reliability-Based Approach of Blood Bank Supply Chain Management with Distribution Center Using Vector-Evaluated Genetic Algorithm

Ajay Singh Yadav
SRM Institute of Science and Technology

Anupam Swami
Government Post Graduate College

Navin Ahlawat, Dhowmya Bhatt and Tripti Pandey
SRM Institute of Science and Technology

CONTENTS

2.1   ROLE OF "MATHEMATICS" IN BLOOD BANK SUPPLY CHAIN INVENTORY MODELS

"Mathematics" has been linked to camp control until the end of time. "Mathematics" is used in the majority areas of everyday existence. Companies make use of "mathematics" in the areas of "accounting, inventory, marketing, revenue forecasting, and financial analysis." "Mathematics" translates to "patience, discipline, and problem-solving." Since "mathematics" is not a recognized science, other than the basis of almost all other subjects, mathematical rules have a very stable purpose. There have been cases in which management decisions have been considered very successful, but when they have been tested on the basis of "mathematics," they have been quite flawed. These save the organizations from jumping into the incorrect car and wasting time and money. Therefore, it is very important that every decision made is accurately measured by mathematical tests. Only then can we ensure that the decision is valid. If a decision is based on the theoretical foundations of an organization and the recognition of mathematical aspects, it should gain the confidence of organization. Some people may think that the solutions provided by mathematical tools are also unsatisfactory because they consider a very ideal set of circumstances which one does not encounter in real life. However, under these situations, it can be argued that assumptions are considered too ideal, although sensitivity analysis provides a way to predict model behavior when circumstances change under ideal conditions.

If compassion psychoanalysis also indicates that the explanation is relatively unwavering, then the clarification can be considered quite accurate and feasible. With the onslaught of nonautomated inventory calculation techniques, we comprise the added improvement of moving away on or after unchanging and unbending assumption. Computerized techniques allow us to choose an acceptable solution instead of finding the optimal solution. Using computerized techniques, we can move away from reality and use satisfactory values for all parameters of the organization. In this way, "mathematics" evolved faster into the reality of noncomputer computing technologies. In view of the rapid progress in all fields of science, day-by-day theories and methods are evolving very fast. The theory related to this is rapidly upgraded. The state of "mathematics" in current affairs is that of the adviser who sets the guiding principle for others. The tools and techniques of this science are evolving little by little, which provide an improved performance space for further hypothetical and practical study expansion. Scientists and researchers want to learn more about the complexity of this science so that newly discovered terrains can serve as a guide for the whole human being. Since the emergence of those who strive to improve the quality of life, "mathematics" has been a very stable and loyal friend.

2.2 BLOOD COLLECTION AND PROCESSING

A blood bank is an institution that "collects," "tests," "processes," and "stores the blood" and its mechanism for future use. The blood bank's main functions are to make arrangement and demonstrate the collection of blood and its system. The main goals of a blood bank are to ensure that there is sufficient blood for patients required for blood transfer and to make sure the waste of blood products is minimized. The blood supply to a particular blood bank comes from expeditions and blood camps. "Donors who come to the bank and donate blood are other sources of supply." Many large blood banks collect their blood from blood banks' shares and donations. Blood is collected and stored in plastic bags that contain anticoagulant solutions. "Blood is collected by donation and stored in plastic bags, often called whole blood (WB)."

A blood bank can also break the blood collected in camps. A single empty bag is used to store WB, whereas a third empty bag is used to collect blood that must be separated into the apparatus. The triple blood bag is a system consisting of a main blood bag and several connected satellite bags to assemble the system. To separate the different mechanisms, a triple/quad blood bag is centrifuged in a high-speed centrifuge. Due to centrifugation, several layers in the main bag diverge according to their density. The lightest component, plasma, is precipitated in the upper layer, followed by platelets in the center and red blood cells (the heavier ones) at the bottom. This mechanism can be emptied from the main bag to their respective satellite bags, when the layers are separated. Since white blood cells in the patient's body can cause many complications, some modern blood bags are equipped with leukocyte reduction filters. These filters eliminate most leukocytes that otherwise differentiate with red blood cells. Individual blood bags are very popular in quadruple pocket blood banks with capacities of 350 and 350 mL, as well as 450 and 500 mL. The main advantage of dividing the WB into plasma is that only the necessary plasma can be provided to the patient.

2.3 GENETIC ALGORITHM

A genetic algorithm was developed by Holl and colleagues in the 1960s and 1970s. Genetic algorithms are based on the theory of evolution, which explains the origin of species. In nature, vulnerable and unsuitable species in their environments are at risk of extinction through general selection. In the language of genetic algorithms, a solution vector $x \in X$ is called an individual or chromosome. Chromosomes have individual units called genes. One or more properties of each gene wheel chromosome. The first implementation of the genetic algorithm by Holl and colleagues assumes that genes are binary numbers. Later implementations introduced different types of genes. Typically, the chromosomal solution space contains a singular solution x. For this, a mapping system between solution space and chromosomes is required. This assignment is called coding. In fact, genetic algorithms work to code a problem, not the problem itself, but a collection of chromosomes, a population. Population is usually randomly initiated. During the research, the population includes both inclusive facilities and installation solutions, which means that they dominate the same solution.

Holland also provided a convergence record (schema set) for the global optimum, which has chromosomal binary vectors. Genetic algorithms use two operators to generate new solutions from existing solutions: crossover and mutation. The crossing operator is the most important operator of the genetic algorithm. At the crossroads, two chromosomes, called parents, are usually combined into new chromosomes, called lineages. The parents are chosen from existing chromosomes in the population, in which fitness is preferred, so that the offspring get good genes, which will make the parents fit. As a result of repeated application of the crossing operator, the population contains a large number of chromosomal genes, which ultimately leads to a globally satisfactory solution for marble industries. The mutation operator introduces random changes in chromosome properties. Mutations are usually applied at the level of the gene. In specific genetic algorithm implementations, the mutation rate (the probability of changing the properties of a gene) is very small and depends on the length of the chromosome. Therefore, the mutated chromosome is not very different from the original one. Mutations play an important role in genetic algorithms.

2.4   LITERATURE REVIEW

Yadav and Swami [1, 2] presented an integrated supply chain model for the degradation of basic products with an adapted linear demand and in a climate of disruption and inflation and a constraint varying in time for a model and the portion size of the female stock. Yadav et al. [3–6] introduced a supply chain warehouse for the expiration of two stocks and inflation and proposals for an inventory model for the deterioration of two stocks and goods with varying costs and deterioration and discussed the analysis of green supply chain inventory management for warehouse storage and environmental collaboration using a genetic algorithm and sustainability performance using a genetic algorithm. Yadav and Kumar [7] demonstrated the management of the supply chain of electronic components for storage in collaboration with the environment and neural networks. Yadav et al. [8–10] examined the effect of inflation on a two-stock commodity stock, which was exacerbated by changing needs and shortages and discussed a model of inventory, which was inflationary for the deterioration of goods in two-inventory systems and proposed an obscure store nonmerchandise model before temporarily deteriorating the goods with a conditional late payment permit. Yadav [11] analyzed supply chain management in optimizing warehouses with logistics using the genetic algorithm. Yadav et al. [12, 13] explained the inventory model for two bearings with optimized soft IT functionality. Yadav [14] explained the modeling and analysis of the supply chain inventory model with two-stage economic transfer problems using the genetic algorithm.

2.5   BLOOD SUPPLY CHAIN

The delivery of a blood collection center is almost identical to the other logistics. The blood is exported to blood storage facilities through a procedure of manufacture and wrapping at a blood center. A shipping technique is organized by the individuality of the individual blood collection sites. Figure 2.1 shows a supply chain process of the blood collection chain.

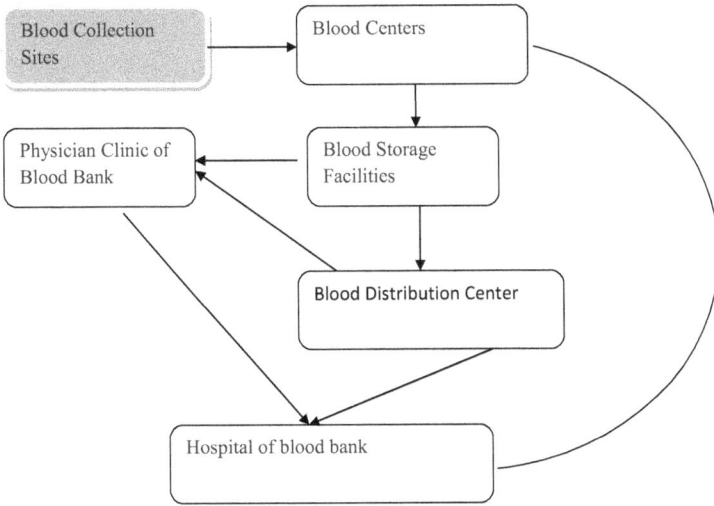

FIGURE 2.1 Blood supply chain.

Blood banks are divided into recommendation types and universal types. A blood bank can buy blood from blood centers and the medical clinic of the blood bank. However, a receipt must be distributed in the form of prescription blood. Blood storage devices communicate blood collection point policy to the blood center or clinic according to the sharing strategy. These last procedures differ according to the types of blood collection sites. In exacting, the distributor should inform the government of the provision and buy and make use of therapeutic blood collection sites. The blood distribution center manages the patient's chemical blood collection sites, such as injections or mixtures. In blood collection stores, prescriptions or generic blood collection points are sold in accordance with the regulations. This process allows blood to be donated to the last hospital in the blood bank.

2.6 "BOOLEAN FUNCTION APPROACH FOR RELIABILITY" OF BLOOD BANKS

In this chapter, the author considers blood banks. The entire organizations consist of six main parts. The first part of the scheme is blood collection sites and is represented by D_1. The second part of the system is the blood center and is represented by D_2, D_3, and D_4. The third part of the system is processing blood storage facilities and homogenizing the blood and is represented by D_5. The fourth part of the system is the blood delivery center and is represented by D_6. The fifth part of the system is the blood bank's physician clinic. This means that the blood is poured into coated paper cartons or plastic bottles. It is sealed and is represented by D_7. The last part of the scheme is the delivery to a hospital of banked blood. This means that blood cans or bottles are kept in defensive delivery containers and kept cool. They are delivered in cooled trailers to warehouses and then sent to individual hospitals, where they are housed in cases of cooling performance and are represented by D_{10}.

2.7 FORMULATION OF THE MODEL

By using Boolean algebra, the condition of efficiency of the successful operations of these blood banks in terms of logic matrix is expressed as under:

$$\text{MATRIX} \quad \begin{vmatrix} D_1 & D_2 & D_3 & D_5 & D_6 & D_7 & D_8 & D_{10} \\ D_1 & D_4 & D_3 & D_5 & D_6 & D_9 & D_8 & D_{10} \end{vmatrix} \tag{2.1}$$

By using laws of algebra of logics, equation (2.1) may be written as:

$$F(D_1, D_2, D_3, \ldots, D_{10})$$

$$= \begin{vmatrix} D_1 & D_3 & D_5 & D_6 & D_8 & D_{10} \end{vmatrix} \cap F \begin{vmatrix} D_2 & D_4 & D_7 & D_9 \end{vmatrix} \tag{2.2}$$

where

$$F(D_1, D_2, D_3, \ldots, D_{10}) = \begin{vmatrix} D_2 & D_7 \\ D_4 & D_9 \end{vmatrix} = \begin{vmatrix} P_1 \\ P_2 \end{vmatrix} \tag{2.3}$$

$$P_1 = \begin{vmatrix} D_2 & D_7 \end{vmatrix} \tag{2.4}$$

$$P_2 = \begin{vmatrix} D_4 & D_9 \end{vmatrix} \tag{2.5}$$

$$P_1' = \begin{vmatrix} D_2' \\ D_2 & D_2' \end{vmatrix} \tag{2.6}$$

$$P_2' = \begin{vmatrix} D_4' \\ D_4 & D_9' \end{vmatrix} \tag{2.7}$$

$$= (D_2' + D_2 D_7')(D_4' + D_4 D_9')$$

Using orthogonalization algorithm, equation (2.3) may be written as:

$$F(D_1, D_2, \ldots, D_{10}) = \begin{vmatrix} P_1 \\ P_1' & P_2 \\ P_1' & P_2' \end{vmatrix} \tag{2.8}$$

$$P_1 = \begin{vmatrix} D_2 & D_7 \end{vmatrix}$$ (2.9)

$$P_1' P_2 = \begin{vmatrix} D_2' \\ D_2 & D_7' \end{vmatrix} \cap \begin{vmatrix} D_4 & D_9 \end{vmatrix}$$

$$= \left(D_2' + D_2 D_7' \right)\left(C_4 C_9 \right)$$ (2.10)

$$= \left(D_2' D_4 D_9 + D_2 D_4 D_7' D_9 \right)$$

$$P_1 P_2' = \begin{vmatrix} D_2' \\ D_2 & D_7' \end{vmatrix} \cap \begin{vmatrix} D_4' \\ D_4 & D_9' \end{vmatrix}$$ (2.11)

$$= \left(D_2' D_4' + D_2' D_4 D_9' + D_2 D_7' D_4' + D_2 D_7' D_4 D_9' \right)$$

Using all these values in equation (2.8), one can obtain:

$$\begin{vmatrix} D_2 & D_7 \\ D_2' & D_4 & D_9 \\ D_2 & D_4 & D_7' & D_9 \\ D_2' & D_4' \\ D_2' & D_4 & D_9' \\ D_2 & D_4' & D_7' \\ D_2 & D_4 & D_7' & D_9 \end{vmatrix}$$ (2.12)

Using this result in equation (2.2), we have:

$$F(C_1, C_3, \ldots, C_{10})$$

$$= \begin{vmatrix} D_1 & D_2 & D_3 & D_5 & D_6 & D_7 & D_8 & D_{10} \\ D_1 & D_2' & D_3 & D_4 & D_5 & D_6 & D_8 & D_9 & D_{10} \\ D_1 & D_2 & D_3 & D_4 & D_5 & D_6 & D_7' & D_8 & D_9 & D_{10} \\ D_1 & D_2' & D_3 & D_4' & D_5 & D_6 & D_8 & D_{10} \\ D_1 & D_2' & D_3 & D_4 & D_5 & D_6 & D_8 & D_9' & D_{10} \\ D_1 & D_2 & D_3 & D_4 & D_5 & D_6' & D_7' & D_8 & D_{10} \\ D_1 & D_2 & D_3 & D_4 & D_5 & D_6 & D_7' & D_8 & D_9 & D_{10} \end{vmatrix}$$ (2.13)

Since RHS of equation (2.12) is the disjunction, the reliability of considered blood banks is given by:

$$R_S = P_r(f(R_1, R_2, R_3, \ldots, R_{10}) = 1)$$

$$= \begin{bmatrix} U_1 U_2 U_3 U_5 U_6 U_7 U_8 U_{10} + \\ U_1 U_2' U_3 U_4 U_5 U_6 U_8 U_9 U_{10} + \\ U_1 U_2 U_3 U_4 U_5 U_6 U_7' U_8 U_9 U_{10} + \\ U_1 U_2' U_3 U_4' U_5 U_6 U_8 U_{10} + \\ U_1 U_2' U_3 U_4 U_5 U_6 U_8 U_9' U_{10} + \\ U_1 U_2 U_3 U_4' U_5 U_6 U_7' U_8 U_{10} + \\ U_1 U_2 U_3 U_4 U_5 U_6 U_7' U_8 U_9' U_{10} \end{bmatrix}$$

$$= \begin{bmatrix} U_1 U_3 U_5 U_6 U_8 U_{10} \begin{Bmatrix} U_2 U_7 + \\ U_2' U_4 U_9 + \\ U_2 U_4 U_7' U_9 + \\ U_2' U_4' + \\ U_2' U_4 U_9' + \\ U_2 U_4' U_5 U_7' + \\ U_2 U_4 U_7' U_9' \end{Bmatrix} \end{bmatrix}$$

$$= \begin{bmatrix} \{U_1 U_3 U_5 U_6 U_8 U_{10}\} \begin{Bmatrix} U_2 U_7 + \\ (1 - U_2) U_4 U_9 + \\ U_2 U_4 (1 - U_7) U_9 + \\ (1 - U_2)(1 - U_4) + \\ (1 - U_2) U_4 (1 - U_9) + \\ U_2 (1 - U_4) \end{Bmatrix} \begin{Bmatrix} U_5 (1 - U_7) + \\ U_2 U_4 (1 - U_7)(1 - U_9) \end{Bmatrix} \end{bmatrix}$$

The appearance for M.T.T.F in this case is given by:

$$\text{M.T.T.F.} = \int G_{SE}(t)\,dt$$

$$= 1/a\ (3/10 - 5/9 + 4/8 - 2/7 + 1/6)$$

$$= .1763072/a$$

2.8 SIMULATION

A simulation blood collection site is to compare and verify the list management of the blood storage facilities in the supply chain. A virtual system comprises ten blood centers, bulk, ten blood distribution centers, and 55 blood collection sites stores. Table 2.1 shows 55 blood collection sites parameters for simulation.

The simulation of the supply chain was approved out with each organization method over 700 near days. We obtained results by calculating the sum of the 5000 executions of the simulations. To evaluate each method of managing blood stocks, we contrast sale prices, sale account, order number, delivery cost, share price, and net profit. Equation 2.10 describes the calculation of distribution costs, share price, and net income (Table 2.2).

$$\textbf{Blood Bank delivery Cost} = \int_1^{10} \left\{ \begin{matrix} \text{blood sales account}_1 \times \text{blood account}_1 \times 1.29 \\ + \text{packaging disposal cost} \end{matrix} \right\}$$

TABLE 2.1
Product Parameter Blood Bank Supply Chain

Product	Max BIC	Minimum Required BIC	Initial BI	Price	Demand (%)
BBC1	8.0	2.0	13	216	45
BBC2	7.9	2.1	21	222	55
BBC3	7.8	2.9	31	223	65
BBC4	7.6	2.8	41	224	17
BBC5	7.9	2.7	51	225	47
BBC6	7.7	2.6	61	226	53
BBC7	7.6	2.5	71	227	57
BBC8	6.5	2.1	81	228	96
BBC9	6.6	2.2	23	221	87
BBC10	6.1	2.7	12	242	75

TABLE 2.2

Simulation Result Blood Bank Supply Chain

Product	Blood Sale Price	Blood Sale Account	Blood Order Count	Blood Delivery Cost	Blood Stock Price	Blood Net Profit
VOQBB	931	742	735	119	776	225
TOQBB	932	741	732	119	923	229
DOQBB	933	741	736	118	840	228
VEGA	938	741	731	119	665	226
Total average	934	741	737	119	851	224

$$\textbf{Blood Bank stock Cost} = \int_{1}^{10} \left\{ \begin{array}{l} \text{current blood inventory quantity}_1 \times 1.798 \\ + \text{packaging disposal cost} \end{array} \right\}$$

$$\textbf{Blood Bank net Cost} = \int_{1}^{10} \left\{ \begin{array}{l} \text{blood sale cost} \times 1.23 - \text{blood delivery cost} \\ - \text{blood stock cost} + \text{packing disposal cost} \end{array} \right\}$$

2.9   CONCLUSION

This chapter studies the effectual method of managing a list of blood storage facilities in the supply chain of blood collection centers. We study the supply chain of blood collection site and take out modeling and simulations. We have a residential supply chain network optimization model to run the "delivery, testing, processing, and distribution of a perishable product based on human blood." The unique aid in this chapter includes the operational model of blood chain management, which presents the extraordinary features of capturing the feasibility of this product that saves lives from side to side with the use of curve multipliers. It includes the costs of squander disposal/disposal. It assesses the costs of bottlenecks and surplus at demand points and quantifies supply-side supply risk. A list policy is an important factor in determining the order time and quantity. It is also important to manage the optimal benefits of the supply chain. Therefore, to increase profits, a trade-off between consumption and control has to be reduced. This letter proposes list strategies using the vector-evaluated genetic algorithm (VEGA). The proposed VEGA calculates the optimal order from the existing stock at the expected standard time. We compare blood bank order quantity (VOQBB), blood bank order (TOQBB) volume, blood bank disposal order (DOQBB) quantity, bank order quantity of bank of blood (EQEBB), and the VEGA. The results of the simulations show the effectiveness of the orders related to the remaining orders and the order quantity specified. The algorithm for the VEGA fulfills both conditions. The supply chain of the blood collection site is a useful method for managing list rules for blood storage facilities. The limitations of this study are as follows. It is difficult to think of the number of blood distribution centers and blood. Apart from this, we did not reflect the features of the demand.

REFERENCES

1. Hsieh, T.P., Dye, C.Y. and Ouyang, L.Y. (2008): Determining optimal lot size for a two-warehouse system with deterioration and shortages using net present value. *European Journal of Operational Research*, 191 (1), 180–190.
2. Benkherouf, L. (1997): A deterministic order level inventory model for deteriorating items with two storage facilities. *International Journal of Production Economics*, 48 (2), 167–175.
3. Yadav, A.S. and Swami, A. (2018): A partial backlogging production-inventory lot-size model with time-varying holding cost and Weibull deterioration. *International Journal Procurement Management*, 11 (5), 639–649.
4. Yadav, A.S. and Swami, A. (2018): Integrated supply chain model for deteriorating items with linear stock dependent demand under imprecise and inflationary environment. *International Journal Procurement Management*, 11 (6), 684.
5. Bhunia, A.K. and Maiti, M. (1998): A two-warehouse inventory model for deteriorating items with a linear trend in demand and shortages. *Journal of the Operational Research Society*, 49 (3), 287–292.
6. Goswami, A. and Chaudhuri, K.S. (1992): An economic order quantity model for items with two levels of storage for a linear trend in demand. *Journal of the Operational Research Society*, 43, 157–167.
7. Lee, C.C. (2006): Two-warehouse inventory model with deterioration under FIFO dispatching policy. *European Journal of Operational Research*, 174 (2), 861–873.
8. Sahooa, L., Bhuniab, A.K. and Kapur, P.K (2012): Genetic algorithm based multi-objective reliability optimization in interval environment. *Computers & Industrial Engineering*, 62, 152–160.
9. Yadav, A.S. and Swami, A. (2019): A volume flexible two-warehouse model with fluctuating demand and holding cost under inflation. *International Journal Procurement Management,* 12 (4), 441.
10. Yadav, A.S. and Swami, A. (2019): An inventory model for non-instantaneous deteriorating items with variable holding cost under two-storage. *International Journal Procurement Management*, 12 (6), 690.
11. Yang, H.L. (2004): Two warehouse inventory models for deteriorating items with shortages under inflation. *European Journal of Operational Research*, 157, 344–356.
12. Zhou, Y.W. and Yang, S.L. (2005): A two-warehouse inventory model for items with stock-level-dependent demand rate. *International Journal of Production Economics*, 95 (2), 215–228.
13. Yadav, A.S., Taygi, B., Sharma, S. and Swami, A. (2017): Effect of inflation on a two-warehouse inventory model for deteriorating items with time varying demand and shortages. *International Journal Procurement Management*, 10 (6), 761.
14. Yadav, A.S. (2017): Modeling and analysis of supply chain inventory model with two-warehouses and economic load dispatch problem using genetic algorithm. *International Journal of Engineering & Technology (IJET)*, 9 (1), 33–44.

3 Synthesis and Design of Energy- and Power-Efficient IoT-Enabled Smart Park

Praveen Kumar Malik
Lovely Professional University

CONTENTS

3.1 INTRODUCTION

Recently, the Internet of Things has been applied from numerous points of view. The shrewd stopping framework is one portion of the innovation of the Internet of Things. The idea of the Internet of Things begins from a gadget that can be followed, controlled, or checked over the web. One of the frameworks of savvy stopping is to know the state of the parking area by means of the web. This is identified with stopping issues, one of which is the trouble of knowing the state of empty space in the wide parking area so that the driver invests his energy just to discover a stopping place alongside the larger number of vehicles [1–2]. Issues identified with stopping can be fathomed if the driver can be educated in advance about the accessibility of parking spots around the ideal goal. As a result, the idea of the Internet of Things applies to the keen stopping framework.

Different methodologies and research have been done to defeat stopping issues. Since the early 1970s, savvy stopping has been executed all through Europe, the United Kingdom, and Japan. The underlying framework is shown in the driver's stopping data, for example, accessibility status and additionally the measure of room accessible. Increasingly intricate savvy stopping fuses further developed innovation to serve clients with various needs [3–4]. At present, there are sure-stopping frameworks that can give continuous data about accessible parking spots. Such frameworks require proficient sensors to be put in parking garages to screen parking spots and quick information handling units to increase reasonable bits of knowledge of information gathered from different sources. As indicated by this information, this examination was to plan a model of the stopping checking framework so as to build up a stopping framework regarding the previous framework. In this framework, we utilized NodeMCU microcontroller, which was upheld by the Wi-Fi framework [5–6]. The proposed savvy stopping framework was actualized utilizing a versatile application that interfaces with the cloud. This framework pushed clients to know the accessibility of parking spots continuously.

Most of the parks that are situated in any of the residential neighborhoods in different cities in India are not automated. They are operated by persons manually. Different facilities such as gardening, lighting, and security-related issues are also operated manually. Very few parks that are at very prominent places of the city have automatic solar lights to light the park at night. It is also observed that wastage of electricity is also there when park lights are used, if they are operated manually. We have observed that lights remain ON during the day time also if they are not managed properly. Especially when the person who has to take care of the lights is absent from work due to any reason or other, the operation of the lights becomes pathetic. There is no foolproof system in any of the social park where the supply of the water is kept under electronics control. A layman simply assumes the moisture and humidity level of the soil before running the motor. However, whenever the layman feels that he should water to the plants or trees, he simply starts the motor and there is no control of the water supply and electricity to the motor. Sometimes, it is observed that the motor runs more than the time required. Even in some of the parks, water storage tank is also there which can store hundreds of liters of water. Usually, there is a layman who operates the motor to fill the tank; it is observed that even if the tank overflows, the motor would be still running. In these types of situations, lot of electrical energy is also wasted [7].

During the summer days, to keep the plants and trees healthy, sprinklers and foggers are used. In manually operated parks, there is no monitoring of running the sprinklers and foggers. It is up to the discretion of the layman or the person who is taking care of the park. Most of the time he forgets to run the sprinklers and foggers at the required time. Sometimes it has even been observed that he runs the sprinklers and foggers more than the time required. Excess water to the plants or trees is also not good for their health. In every park, there is an electrical and electronics control unit from which layman controls the water supply motor, which is connected to the sprinklers and foggers. Sometimes it is observed that, due to short circuit, fire happens in the control room and damages the property and occasionally humans as well. There is no provision in most of the parks where sensors are installed regarding who will inform the authority in case of fire.

In any particular park, especially in a particular colony or residential society, it is observed that persons from other locations also enter and may create some problems. Sometimes, unusual and unwanted persons also enter the park, which creates problems for those who are in the park or those who have the right to use the facilities available in the park. There should be a system that will keep an eye and that will monitor persons entering the park, whether they an authorized person or not. Unwanted miscreants should not be allowed to enter the park. It is also observed that sometimes the number of people who use the park is much more than the total capacity of the park. There should be a system that will monitor the number of persons entering the park. Once the capacity of the park is full, further persons should not be able to enter the park. In most of the parks and the surrounding area, there is no provision for checking the quality of air and its necessary remedy. If the air quality near any park or society is poor, we have to consider many health problems. But if we come to know that the air quality near our surrounding area is not good, we can take some actions to improve it or we can suggest the higher authorities of our city or province take necessary actions. Similarly, carbon monooxide gas monitoring can be also done to improve air quality of our surrounding area. During nighttime or even in day time, it has come to our notice that some persons enter into the park after drinking alcohol, etc. One can easily place the alcohol sensor to monitor the consumption of alcohol while entering into the park (Figure 3.1).

3.2   RELATED WORK

Thombare et al. show that in many cities, we observed wastage of water and electricity and improper use of dust bin. India is one of the largest freshwater users in the world [8]. Automation will make efficient use of electricity as well as water. In this chapter, the author suggests the use of garbage collection using GSM. Light-dependent resistor (LDR) is used to control wastage of electricity. Jorge E. Gómez et al. describe the purpose of the Internet of Things to make permanent and inclusive the access with a great variety of devices connected to the Internet [9].

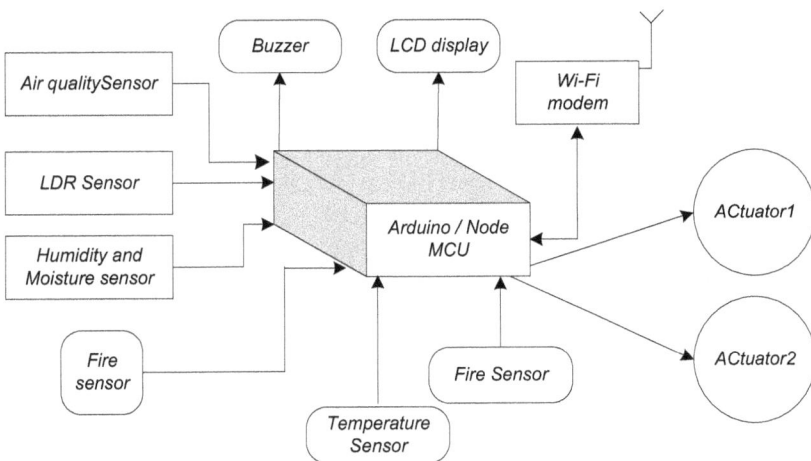

FIGURE 3.1   Architecture of the proposed model.

3.3   PROBLEM FORMULATION

In most of the parks in India, there is no such solution, which is provided in this chapter. Usually, there is a layman or a person who is responsible for all the activities related to the park and surrounding area. For example, to operate the motor and to provide water to the plants and trees in the park, the layman simply switches "ON" the motor and does not care about how much water has been used in the process, and roughly switches "OFF" the motor assuming that water supply to the plants and trees was sufficient. The number of persons entering into the park is not monitored in any of the systems. An automated light system can be found in some of the parks. There is no system that can monitor that the person entering into the park is drunk or not. Nowhere the system is installed which can monitor the fire and/or temperature at the place where the motor or electrical appliances are installed. In the existing solution, the layman does not check the level of moisture of soil to run the motor. Air quality index is also not checked in most of the parks. And if it is checked, there is no action taken to improve the air quality level.

3.4   SYSTEM OVERVIEW

3.4.1   LIMITATION OF EXISTING SOLUTION

Proposed design can only be applicable if and only if the electricity is there in the park and/or area. For more effective working of the solution, Internet is also required. Awareness about the uses of the electronics- and electrical-related devices is also required.

3.4.2   IMPLEMENTATION

We propose the novelties of the proposed work in the form of a book chapter in the following steps:

Design suit will be based on the Arduino-based microcontroller, and it is supporting components and devices especially based on sensors and ESP8266 module which can support the Internet of Things (Figure 3.2).

As stated earlier, also a number of sensors could be used to control the different physical and nonphysical activities in any park or social area. An LDR module could be used and interfaced in this chapter. LDR is a light-dependent component whose characteristics change when the light falls on it. Intuitively when the light falls on the sensor, its resistance decreases; however, when the intensity of the light is less on the LDR, then the resistance of the LDR will be more. We are going to use this property to make the lights of the park to be switched ON or OFF. During the evening time when the intensity of the light will be less, the light of the park will be turned ON automatically. A special module of the program could also be written for this activity.

Interfacing module for LDR sensor can be taken from Figure 3.3. When the light falls on the LDR sensor, it will produce logic 1, and when the intensity of the light is less, it will give logic 0. This logic can be further used to make use of the corresponding relay to make the lights switch ON or OFF. Humidity and soil moisture sensor module could also be used to monitor the humidity and moisture of

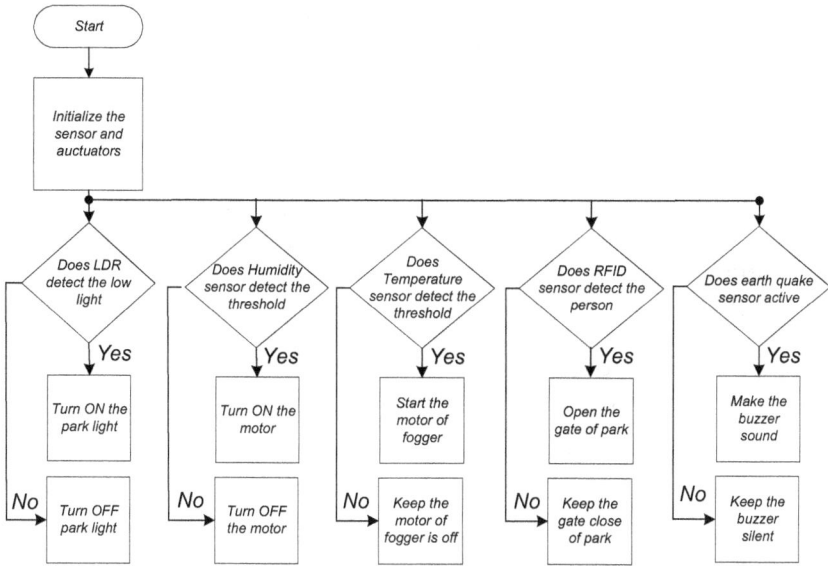

FIGURE 3.2 Block diagram of the proposed model. LDR, light-dependent resistor.

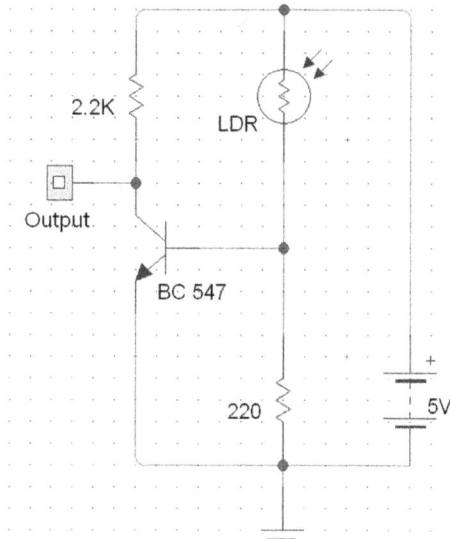

FIGURE 3.3 Interfacing module for LDR sensor. LDR, light-dependent resistor.

the soil of the park. Humidity and soil moisture will be monitored by the sensors and microcontroller unit. If the system finds that the level of the humidity and moisture is less than the threshold value, a relay-connected motor will be turned ON. This motor will remain ON till the level of the humidity and moisture reaches to the desired level. A special module of the program will also be written for this activity. The following circuit can be used to interface the relay to microcontroller.

Temperature sensor module is also proposed in this chapter. The module will be connected to the motor of the sprinklers and foggers of the park. In the summers, when the temperature rises from the set level, a motor will be connected through the relay to run the motor. Figure 3.4 can be further used for the same purpose. This motor will continue until the temperature goes down to the set level. A special module of the program could also be written for this activity. A fire sensor module is also proposed to monitor the status of the fire if any at the place where most of the electrical pieces of equipment are placed in the park. This equipment may be power panel, motor operation board, and so on. Once if the fire is detected, then a buzzer can be used to make the sound such that precaution or action can be taken against the fire. The following circuit can be used to interface the buzzer with the design (Figure 3.5).

Radio frequency identification (RFID)–enabled cards could also be provided to the persons those who are authenticated and legible to the uses of the park.

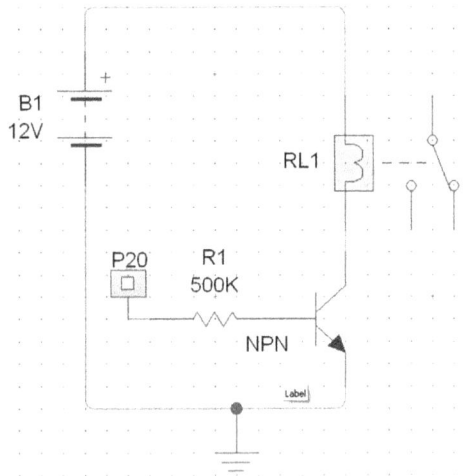

FIGURE 3.4 Interfacing module for relay with microcontroller.

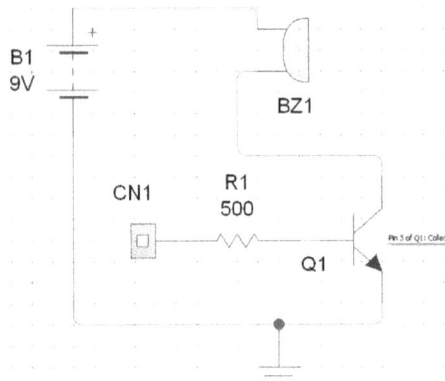

FIGURE 3.5 Interfacing module for buzzer with microcontroller.

There could be two gates in the park: one for entry and other for exit. Once if the authenticate person places the card on a particular location of the sensor at the gate, the gate will open automatically, and after a certain interval of time, the gate will close by itself. This sensor will also keep an eye on the number of persons entering the park. Once if the capacity of the park is full, the gate will not open after placing the card until some person exits from the park. On the second gate, users can exit; when the user places the card on the sensor, automatically the gate will open once the count of the number of persons in the park reduced by 1. These two gates will make a harmony and synergy between them to keep an eye on the number of users within the park. In metro cities and small towns, major casualties happen during the earthquake also. Due to the enhancement of the technology, if we could be in a position to know whether it is an earthquake, it can save the lives. In the proposed chapter, we are planning to incorporate the earthquake sensor that will monitor the status of the earthquake, and if it so. The module will immediately make the sound on the buzzer and will also inform the concerned authority through the Internet of Things.

The air quality sensor module is also incorporated in the proposed concept. This module will monitor the air quality in the atmosphere and will inform the concerned person twice a day. If the quality of the air is getting bad or worse, then concern authority or Municipal Corporation will come to know the situation, and they will take the appropriate action to get it fine.

3.5 INTERFACING OF NUTTYFI/ESP8266 WITH ANALOG SENSOR

The following section would comprise the interfacing of sensors with the Internet using the ESP8266 microcontroller. Analog sensing element senses the external parameters such as temperature, humidity, and so on and offers associate degree along voltage as an output. The output voltage is also within the range of 0–5 V. Logic high is treated as "1" (3.5–5 V), and logic low is indicated by "0" (0 V) (Figure 3.6).

There are differing types of sensors that turn out continuous analog output, and these sensors are analog sensors. This continuous output made by the analog sensors is proportional to the measure. There are numerous styles of analog sensors; sensible samples of numerous styles of analog sensors are as follows: accelerometers, pressure sensors, lightweight sensors, sound sensors, temperature sensors, and so on. The system comprises +12 V power supply, +12 V to +5 V converter, NuttyFi,

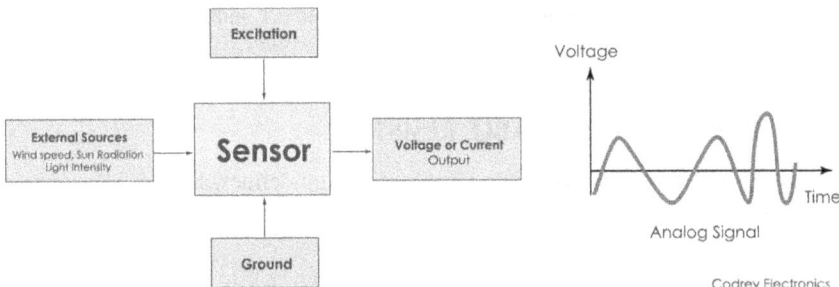

FIGURE 3.6 Analog sensor – block diagram.

TABLE 3.1

Components Used with NuttyFi

S.N	Name of the Components	Quantity	Specifications
	Components Used to Interface ESP8266 with Variable Resistor		
1	+12 V power supply	1	Output: +12 V/1 A
2	+12 V to 5 V converter	1	Output: +5 V/1 A
3	NuttyFi	1	Analog pin 1
			Digital pin 10
			FTDI connector for programmer
4	LCD 20 × 4	1	16-pin LCD with backlight pins
5	POT with breakout board	1	Three pins
6	LED with breakout board	1	One red color LED

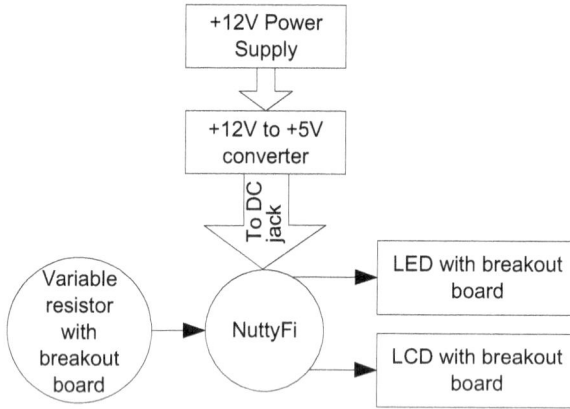

FIGURE 3.7 Block diagram of interfacing of variable resistor with ESP8266.

LED with resistor 330 ohms, POT (potentiometer), and LCD (liquid crystal display). The objective of the project is to display the levels generated by ADC on LCD using NuttyFi by writing the hex file in the flash memory. If the levels increased by certain level, indicator will turn ON to show the status (Table 3.1 and Figure 3.7).

3.6 DESCRIPTION OF INTERFACING OF ESP8266 WITH VARIABLE RESISTOR

The interfacing of the system is as per the given guidelines. Figure 3.8 shows the connection diagram of the system.

1. Connect +12 V power supply output to input of +12 V to +5 V convertor.
2. Connect +5 V output to power up the NuttyFi board, and also connect to other peripheral to power up.

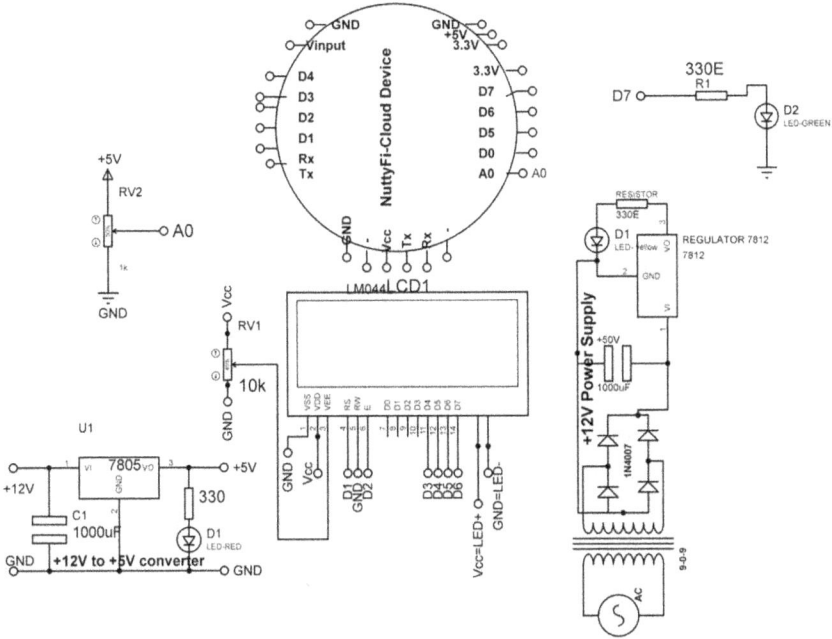

FIGURE 3.8   Circuit diagram of the system.

3. For programming uses, connect FTDI programmer to FTDI connector of the NuttyFi board.
4. Pins 1–16 of the LCD are connected to GND.
5. Pins 2–15 of the LCD are connected to +Vcc to power supply.
6. Two fixed pins of the variable resistor are connected to +5V, and GND of LCD and variable pins of the variable resistor is connected to pin 3 of LCD.
7. E, RS, and RW pins of the LCD are connected to pins D1, GND, and D2 of the NuttyFi.
8. D4–D7 pins of the LCD are connected to pins D3, D4, D5, and D6 of the NuttyFi.
9. Connect +Vcc, GND, and OUT pin of the **POT** to +5V, GND, and A0 pins of the NuttyFi.
10. Connect input and GND pins of the LED breakout board to D7 and GND pins of NuttyFi.

3.7   PROGRAM TO INTERFACE VARIABLE RESISTOR WITH ESP8266

```
#include <LiquidCrystal.h>
LiquidCrystal lcd(D1, D2, D3, D4, D5 and D6); // add library
of LCD
const int POT_Pin=A0; // assign A0 pin to POT
```

```
const int INDICATOR_PIN=D7; // assign D7 as indicator pin
void setup()
{
 pinMode(INDICATOR_PIN, OUTPUT); // set D7 pin as an output
 lcd.begin(20, 4); // initialize LCD
 lcd.print("POT Control..."); // print string on LCD
}
void loop()
{
 int POT_Pin_LEVEL = digitalRead(POT_Pin);// Read POT pin
 lcd.setCursor(0, 1); // set cursor of LCD
 lcd.print("ACTUAL_LEVEL:"); //print string on LCD
 if (POT_Pin_LEVEL >= 240) // compare level
 {
 lcd.setCursor(0, 3); // set cursor of LCD
 lcd.print("LEVEL_EXCEED "); // print string on LCD
 digitalWrite(INDICATOR_PIN, HIGH); // set D7 to HIGH
 delay(20); // wait for 20mSec
 }
 else
 {
 lcd.setCursor(0, 3); // set cursor of LCD
 lcd.print("LEVEL NORMAL "); // print string on LCD
 digitalWrite(INDICATOR_PIN, LOW); // set D7 to HIGH
 delay(20); // wait for 20mSec
 }
}
```

3.8 CONCLUSION

Our proposed solution is better than any of the existing system available in the following terms. As some lights are there in the park which can facilitate the users during the night walk or for park use, in the proposed work, lights of the park can be automated. Manual operation of the ON and OFF of lights is not needed. Whenever the natural light goes down, automatically park lights will be turned ON. These lights will be turned OFF in the morning automatically when the sunlight will be there, i.e., lights are not time dependent, they are natural light dependent. In this chapter, checking of the moisture and humidity of the soil will not be there. The system will continuously monitor the requirement of the water supply to the plants and trees depending upon the moisture and humidity of the soil and atmosphere. When there is necessity for water, it will run the motor automatically. A new concept of the earthquake intimation is also introduced in this chapter. We will use an earthquake sensor in our design. The sensor will continuously monitor the activity of an earthquake. This facility is not provided in any of the parks available. Counting the number of persons in a park will prove useful, which will keep an eye on the number of persons in the park such that it should not be overcrowded.

Air quality monitoring will be there whenever the system finds that air quality is below the threshold level. It will inform the concerned person through SMS or phone. In the proposed work, fire sensor module will be there to monitor the fire status in and around the park, and if so, the system will make a sound through the buzzer and will inform the concerned person through SMS or phone. RFID card–based entry will be there in the park. These RFID cards will be distributed to those persons who are authenticated to use the park amenities facilities.

REFERENCES

1. Lee C, Han Y, Jeon S, Seo D and Jung I 2016 "Smart parking system for Internet of Things" *IEEE Int. Conf. Consum. Electron.* pp. 263–264.
2. Khanna A and Anand R 2016 "IoT based smart parking system" In *Proceedings of International Conference on Internet of Things and Applications (IOTA)*, Pune, India, pp. 266–270.
3. Basavaraju S R 2015 "Automatic smart parking system using internet of things (IoT)" *Int. J. Sci. Res. Publ.* 5 (12) pp. 629–632.
4. Ibrahim F, Jadhav P N, Bandarkar S P, Kulkarni O P and Shardoor N B 2016 "Smart parking system based on embedded system and sensor network" *Int. J. Comput. Appl.* 140 (12) pp. 975–8887.
5. Chinrungrueng J, Sunantachaikul U and Triamlumlerd S 2007 "Smart parking: An application of optical wireless sensor network" In *2007 International Symposium on Applications and the Internet – Workshops (SAINT 2007 Workshops)*, 15–19 January 2007, Hiroshima, Japan, pp. 66–66.
6. Rodier C J, Shaheen S and Kemmerer C 2008 "Smart parking management field test: A bay area rapid transit (BART) district parking demonstration final report" *Transp. Res. Rec.* 2038 (1) pp. 62–68.
7. Septriyaningrum I A, Nugrahadi D T and Ridwan I 2016 "Perancangan Dan Pengembangan Prototype Sistem Parkir" *J. Ilmu Komput.* 3 (2) pp. 146–155.
8. Thombare R B, Pranita S, Swati D and Swati P 2018 "Smart public garden by using Arduino" In *13 International Conference on Recent Innnovation in Science, Engineering and Management.* ICRISEM 2018, pp. 679–683.
9. Gómez J E, Marcillo F R, Triana F L, Gallo V T, Oviedo B W and Hernándeza V L 2017 "IoT for environmental variables in urban areas" Proc. Comput. Sci. 109 pp. 67–74.

4 Machine Learning Security Allocation in IoT using Raspberry Pi

P. Karthika and P. Vidhya Saraswathi
Kalasalingam Academy of Research and Education

CONTENTS

4.1 INTRODUCTION

Internet of Things (IoT) is a system of associated procedure and frameworks to trade or collect information and data produced by clients of implanted [24] sensors in the physical items. Among the protection, vitality, mindfulness, condition, and different concerns, security assumes an imperative job, as the (conceivably touchy) information is sent among different gadgets and various clients. In situations where such information is captured and utilized for non expected purposes, it might prompt the extreme harms of the profitable framework or potentially natural resources. There exist various overviews identified with the IoT [23] security, security of the IoT structures, or explicit parts of the IoT frameworks. We suggest a far-reaching location display to security administration in the IoT frameworks.

We base our proposition on the space display for the information system security risk management (ISSRM) – in this manner, we center on security dangers to the data and information overseen in the IoT framework. The IoT [9–11] frameworks much rely upon the cloud and Internet processing, which makes us think about how vulnerabilities and their countermeasures in the open Web application security project (OWASP) can help while recognizing and dealing with security hazards in the IoT frameworks. To apply accessible arrangements in verifying Web application to the IoT framework, we recommend self-created IoT [23] security suggestion model based on IoT and SRM domain models.

4.1.1 INTERNET OF THINGS SECURITY CHALLENGES IN APPLICATION LAYER

Most important security challenges in the application layer can be obtained by the following:

i. **Authentication**: It allows integration of various IoT procedures from various settings. Amid the confirmation procedure, it is required to verify directing friends utilized for information exchanging/trading alongside inception information hub verification. Therefore, we confirm that the key organization and administration are position difficulties in IoT validation. In this of hubs is constantly essential for a decent and secure [21] execution as it gives counteractive action alongside illicit hub entrance. Validating the start to finish correspondence makes it secure, and the aggressor will not almost certainly parody the messages between these ends. One more test is to keep up verification plans both unimportant and solid.

ii. **Authorization**: It addresses the security [12] problems of illegal hub entrance by determining the entrance rights to various assets. The information ought to be safely ensured and available just to approved clients. As each IoT hub underpins restricted systems for access confirmation, it is continually testing to give suitable approval instrument, which will be helpful and endure to hubs with various capacities.

iii. **Exhaustion of resources**: One of the difficulties is because of assaults, which point is to deplete the vitality of the objective IoT hub through directing circles or broadening bundles exchanging ways. Every hub can be physically assaulted and bargained; thusly, it very well may be utilized to perform pernicious exercises inside the system. In this way, the principal challenge happens when one of the hubs is undermined, and sub sequently, the cryptographic systems are not any more helpful as the aggressor gets an entrance to the keys and encryption conspires through traded-off hub.

iv. **Trust establishment**: It is a critical part of each IoT framework in light of the fact that such frameworks regularly manage touchy information. They store, exchange, repetitively process recover, and take choices upon this sort of information. In such cases, implementing trust assumes to be key elements of the utilitarian conducting all things considered: conventions, components, and elements of the IoT framework. Giving a trust show on application level so as to pursue the rules of information uprightness and privacy sets up a ground test for touchy information security.

4.1.2 Artificial Intelligence Using Machine Learning IoT

The expectations of home security look bright to development in artificial intelligence (AI) (Figure 4.1).

AI is acquainted with more aspects of our lives. The vast majority as of now conveys advanced cells that have AI-fueled voice recognition applications, and driverless vehicles will be a common presence on the road within the next few years. With regard to the home, we additionally have an assortment of brilliant tech that can make our lives simpler. Notwithstanding that, shrewd home frameworks can be utilized to make the home increasingly [21] secure. Increasingly more security organizations are hoping to procure a developer to create IoT arrangements that assist and ensure your home's safety remotely.

4.1.3 Artificial Intelligence and Machine Learning for Video Surveillance

AI for video observation uses computer software programs that break down the pictures from video reconnaissance cameras so as to perceive people, vehicles, or articles. Security [12] contract workers program is the product to characterize

FIGURE 4.1   Artificial intelligence home security looks.

confined regions inside the camera's view (for example, a fenced off zone, a parking garage yet not the walkway, or an open road outside the part) and program for times of day(for example, after the end of business) for the property being ensured by the camera reconnaissance. The "AI" sends a caution in the event that distinguishes a trespasser breaking the "rule" set that no individual is permitted around there amid that season of the day.

The AI program works by utilizing machine vision. Machine vision is a progression of algorithms, or mathematical procedures, which work like a stream outline or arrangement of inquiries to contrast the item observed and a huge number [9, 10] of put-away reference images of people in various stances, points, positions, and developments. The AI inquires as to whether the watched [14] article moves like the reference pictures, regardless of whether it is around a similar size stature in respect to width, on the off chance that it has the trademark two arms and two legs, on the off chance that it moves with comparative speed, and on the off chance that it is vertical rather than even. Numerous different inquiries are conceivable, for example, how much the item is intelligent, how much it is unfaltering or vibrating, and the smoothness with which it moves. Consolidating the majority of qualities from the different inquiries, a general positioning is inferred which gives the AI the likelihood that the item is anything but a human. On the off chance that the esteem surpasses a breaking point that is set, the alarm is sent. It is normal for such projects that they are self-figuring out how to some extent, learning, for instance, that people or vehicles seem greater in specific [6] segments of the observed picture – those regions close to the camera – than in different segments, those being the territories most distant from the camera. In addition to the simple rule restricting humans or vehicles from specific territories at specific occasions of day, progressively complex standards can be set. The client of the framework may wish to know whether vehicles drive one way yet not the other. Clients may wish to realize that there are in excess of a specific preset number of individuals inside a specific region. The AI is fit for keeping up observation of many cameras at the same time. Its capacity to detect a trespasser out there or in downpour or glare is better than people's capacity to do as such.

This type of AI for security is known as "rulebased" on the grounds that a human software engineer must set standards for everything for which the client wishes to be alarmed. This is the most pervasive type of AI for security. Numerous video observation camera frameworks today incorporate this kind of AI ability. The hard drive that houses the program can either be situated in the cameras themselves or can be in a different gadget that gets the contribution from the cameras.

A more up-to-date, non–rule-based type of AI for security called "conduct investigation" has been created. This product is completely self-learning with no underlying programming contribution by the client or security temporary worker. In this sort of investigation, the AI realizes what is typical conduct for individuals, vehicles, machines, and the earth dependent on its own perception of examples of different attributes, for example, measure, speed, reflectivity, shading, gathering, vertical or flat introduction, and so on. The AI standardizes the visual information, implying that it groups and labels the items and examples it watches, developing persistently refined meanings of what is typical or normal conduct for the different objects watched.

4.1.4 How Can Improve Your Artificial Intelligence Security System

With conventional home security gadgets, a [8] framework can be modified to trigger a foreordained reaction dependent on specific occasions. For instance, if the framework is outfitted, it would begin a commencement when an entryway is opened. On the off chance that you neglect to enter the regulations on the keypad inside the time period, it would alarm the police or an off-site security expert to the possible interloper.

While the framework is possible for progress, a few issues could emerge. The greatest issue is that it could prompt an issue with false alerts. False cautions are clearly awkward; however, they can prompt an absence of reaction in case of a genuine security issue. On the off chance that the police or security administrations get such a large number of false cautions inside a brief timeframe, they may choose that future occasions do not warrant a similar desperation of reaction.

The equivalent can be valid for CCTV cameras. Your standard surveillance camera is for the most part an aloof measure. It can go about as an obstruction, yet on the off chance that a wrongdoing happens, they just act to [10,14] record proof; they do not successfully stop the wrongdoing. On account of AI programming it is beginning to modify with the security arrangement of things to come, each part of your home assurance framework will be associated throughout the IoT. The whole lot from CCTV cameras, entryway locks, illumination, and sound [24] sensors will gather information that would then be able to be examined utilizing AI.

By investigating this information, the framework can find out about the propensities and exercises that are normal to the property. When the framework has a profile of normal movement, it would then be able to utilize what it has figured out to deal with the home security development in a progressively viable way. Your determination starts to observe things like locks which not just identify the typical occasions that diverse individuals go [18] back and forth; however, that can likewise incorporate with the cameras to recognize different guests. This will enable the program to settle on educated access choices. The framework might likewise see that you neglected to bolt your entryways when you went out and send a suggestion to your cell phone. When you get the notice, you could utilize the connected cell phone application to bolt the entryways without returning home.

4.1.5 Smart Security Cameras

Surveillance cameras are broadly perceived as a standout among the best robbery obstructions. At the point when the thief observes cameras, the thief is bound to pick another objective. The viability of these cameras has completed them progressively normal in home security. Their expansion in ubiquity is likewise because of the way that we have as of late observed some critical proceeding in [19] CCTV innovation.

Highlighting similar to HD recording, distributed storage space and remote checking are on the whole normal among present-day surveillance cameras. Notwithstanding that, some more up-to-date frameworks have started to utilize [24] AI programming, making CCTV progressively compelling. With AI, a security framework can figure out how to perceive the general populations that have a place in [20] home. These perceptions help the framework when it needs to choose whether or not to generate an alert.

Past facial acknowledgment, surveillance cameras can likewise be customized to perceive objects. To be valuable for selection, the framework differentiates between a possibly compromising outsider and somebody who is moving toward the home to convey a bundle.

4.1.6 SMART LOCKS FRAMEWORK WITH MACHINE LEARNING

The older lock is key to the conventional [6, 15] methods for overseeing access control in the home. In the event that individuals have a key, they can go back and forth however they see fit. With a smart lock, the physical key is a relic of days gone by [22]. Rather, individuals can get entrance utilizing effects like PIN codes and cell phone and machine learning applications. With a smart lock, you can oversee diverse dimensions of various accesses for individuals, and the framework even ventures to such an extreme as to keep up an entrance record.

A significant number of these IoT frameworks likewise accompany highlight intended to build their comfort. While the smart locks are associated with home Wi-Fi, they can modify to open as approaching the front entryway. With some smart locks, the capacity to utilize a computerized individual collaborator to open the entryway utilizing voice directions. At the point when a visitor arrives, you could basically advise your advanced partner to open the front entryway.

The issue with customary IoT security frameworks comes behind to programming. While set up to react to occasion-based triggers, despite everything, they require a human to perform huge numbers of the most critical undertakings. With the advancement of [27] AI, there is to a lesser extent a requirement for human security experts. The smart framework will probably screen itself and even turn out to have the capacity to do things like reaching crisis administrations.

4.1.7 RASPBERRY PI

The Raspberry Pi is a progression of small single-board PCs produced in the United Kingdom by the Raspberry Pi establishment to advance educating of fundamental software engineering in schools and in creating countries. The opening model wound up undeniably better known than foreseen selling outside its objective market for utilization. It does exclude peripherals (for example, keyboards and mice) and cases. However, a few adornments have been included into a few authorized and informal packs.

The association following the Raspberry Pi comprises of two missiles. The initial two models were created by the Raspberry Pi establishment. After the Pi model B was discharged, the establishment set up Raspberry Pi trading, with Eben Upton as CEO, to build up the third model, the B+. Raspberry Pi operation is in charge of building up the innovation, while the foundation is an instructive philanthropy to proceed the humanizing of original software engineering in schools and in creating countries. According to the Raspberry Pi establishment, more than 5 million Raspberry Pi units were sold by February 2015, making it the best-selling British computer. They had sold 11 million units by November 2016 and 12.5 million by March 2017, making it the third best-selling "general purpose computer." In July 2017, sales reached nearly 15 million. In March 2018, sales reached 19 million.

FIGURE 4.2   Raspberry Pi 3 model B | the Pi Hut.

4.1.8   RASPBERRY PI MODELS AND FUNDAMENTAL PC

Raspberry Pi is a minimal effort, fundamental PC that was initially expected to help computing enthusiasm for processing among school-matured kids. The Raspberry Pi is contained on a solitary circuit board and highlights ports for HDMI, USB 2.0, and composite video (Figure 4.2).

Raspberry Pi 3 can run the full scope of ARM GNU/Linux conveyances, counting Snappy Ubuntu Core, just as Microsoft Windows 10. This has an indistinguishable structure factor to the current Raspberry Pi 2 model B; however, it figures out how to collect in both the original BCM2837 and a full 1GB of SDRAM. It additionally includes Wi-Fi and Bluetooth small energy capacities to improve the usefulness and the capacity to control all the more dominant gadgets over the USB ports.

4.1.9   FEATURES OF RASPBERRY PI

- Broadcom BCM2837 64-bit quad core processor fueled single-board computer running at 1.2GHz 1GB RAM
- BCM43143 Wi-Fi ready
- Bluetooth low energy (BLE) on panel
- 40-pin expanded GPIO
- 4×USB 2 ports
- 4-shaft stereo output and composite video port and full estimate HDMI
- CSI camera port for associating the Raspberry Pi camera
- DSI show port for associating the Raspberry Pi contact screen show
- Micro SD port for stacking your working framework and putting away information
- Upgraded exchanged micro USB control source (presently bolsters up to 2.5A)
- Raspberry Pi 3 B+/3B SATA HDD/SSD storage space expansion board,X820 V3.0 USB 3.0 mobile hard disk module compatible with 2.5-inch SATA HDD/SSD/Raspberry Pi 3 model B+ (B plus)/3 B/ROCK64/Tinker board (Figures 4.3 and 4.4).

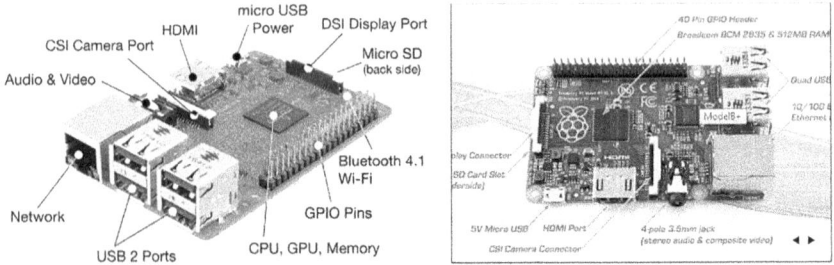

FIGURE 4.3   Raspberry Pi model B+ (old model – board only).

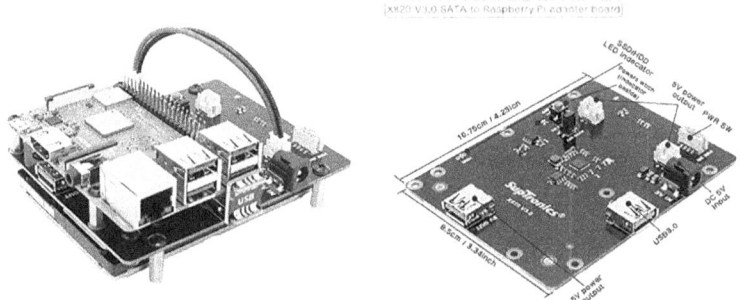

FIGURE 4.4   SATA HDD/SSD/Raspberry Pi 3 model B+ (B plus)/3 B/ROCK64/Tinker board.

4.1.10   How Raspberry Pi Can Be Used for Machine Learning Applications

Figure 4.5 Raspberry pi Model B+ (B Plus)/3 B/ROCK64/Tinker board, little chip-based PC.

Raspberry Pi, a little chip-based PC which is first acquainted with school under-studies, has progressed significantly from that point forward. It has now advanced to performing errands such as face acknowledgment and article acknowledgment. Despite being a solitary board PC, it is valuable in manners that are ground-breaking and functional and are recognizable to any "Linux-head" (Figure 4.5).

4.1.11   Raspberry Pi Meets Machine Learning

Here are few cases where Raspberry Pi has been utilized in combination with different assets for ML applications:

4.1.11.1   Tensor Flow

Raspberry Pi, when utilized with a blend of different assets such as Tensor Flow and Python, has an assortment of utilization. Tensor Flow, an open-source programming library for information stream, can likewise be utilized to effort with Raspberry Pi. It has a superior mathematical calculation inclination and is created by analysts and specialists; a application of AI-based utilize this ML and profound learning.

FIGURE 4.5   Raspberry Pi model B+ (B plus)/3 B/ROCK64/Tinker board, little chip-based PC.

Utilizing the Tensor Flow with Raspberry Pi, request like article discovery in recordings is conceivable. Distinguishing whether there is a pooch in the video or a residence, to check there is a leaving territory accessible before your work environment, and assembling your own vehicle beginning unit or structure a robot that accepts the greatest play to create in a card amusement are all conceivable. A different application is oneself driving car can be worked at an extremely modest cost utilizing Tensor Flow.

4.1.11.2   Google AIY Kit

Google's pack AI is one of the stages to present AI on the Raspberry Pi chip. The first quantity gave an approach to a voice communication utilizing the Raspberry Pi. It gives you a chance to assemble the language processor and interface to be housed in a little box controlled by the Raspberry Pi. The second AI venture was likewise founded on Raspberry Pi; however, instead of the voice, they had a dream unit based around a Pi Bonnet, which depends on the Raspberry Pi. AI declared a pack that had a mix of voice and vision, and it incorporated a Raspberry Pi which is another adaptation of the Raspberry Pi and a flat ground-breaking one.

4.1.11.3   Image Sorting

Great deals of other mind-boggling, huge-scale applications are conceivable. For instance, an implanted framework creator from an industry named Makoto Koike built up a cucumber-arranging framework for utilizing Raspberry Pi 3 as the fundamental organizer to take pictures of the cucumbers. The pictures are caught, and while the Raspberry Pi has the capacity to run profound [25] neural systems, it scuttles a little range neural [25] system on Tensor Flow to order regardless of the picture of a cucumber. A Linux system scuttles more elevated amounts of extra arrangement. An Arduino Micro controls the transport and servo engines and a variety the cucumbers as per the arrangement nourished. A Raspberry Pi chip, alongside Tensor Flow, is utilized to perceive cucumbers as they travel along the transport and sends photographs for further preparing.

4.1.11.4   Others

Space travelers through a virtual copilot utilizing Raspberry Pi, an AI framework, called the sunlight-based pilot guard, to anticipate air ship crashes. The undertaking was finished fruitful with the help of Raspberry Pi. Aurora alert, music hyper disclosure, and discourse to content area portion of the other AI applications. Open CV is an open-source processor visualization library for the utilization of AI, meant to help the utilization of machines in business items; likewise, it has request like article recognition and face discovery, when embraced with Raspberry Pi.

4.1.12 WHERE DO THEY LACK?

Regardless of having huge number of uses, Raspberry Pi cannot contend with increasingly advanced top of the line AI applications. It does not have the correct equipment for overwhelming computational procedures, since it has an inadequate RAM and workstation speed. That is the reason it is important to prepare the system on a PC or a work area and afterward utilize this prepared neural [25] system to the Raspberry Pi, in light of the fact that these requests interest for a great deal of assets which are not finished accessible by the Raspberry Pi. Along these lines bigger, more profound neural systems cannot be utilized with it. We are constantly constrained to utilize fewer processor impressions when computations are increasingly complex. The reasonableness of these PCs may fluctuate, contingent upon the reason; in any case, it tends to be adjusted to suit the necessity.

4.2 EXISTING TECHNOLOGIES AND ITS REVIEW

An IoT multilayer framework enhances the security of multimodal [25–27] biometric data using a combination of watermarking and steganography. It uses a watermark technique to cover faces in photographs of [1, 3] Eigen fingerprints. After that, a technique of steganography is used to hide data as a result of the footprint or the face on an arbitrary of any significance to the biometrics or forensic image watermarked. Recessed mounting locations are set randomly in all three colors of arbitrary image channels.

Biometrics-based personal identification techniques used digital watermarking techniques to incorporate sensitive information like the logo of the company on the host data to protect the rights of intellectual property of the data. They are also used for authentication of multimedia data. Encryption can be applied to the biometric templates to enhance security model which can be in (1) a database, (2) a brand like the smart card, or (3) a biometric device such as a mobile phone with fingerprint [24] sensor that can be encrypted after registration. Then, during authentication, encrypted templates can be read and used to generate the result corresponding with biometric data online. As encryption models are provided, they may not be used or modified without decryption with the appropriate key (Figure 4.6).

background image   foreground image

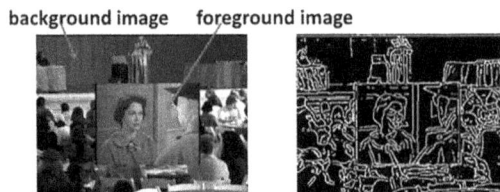

FIGURE 4.6 Sample of image edge detection. (a) Pi P image. (b) Edge image.

The video identification is chiefly centered on recognizing frame level. Furthermore, picture discovery can be demonstrated as recognition into an edge picture for possible square shapes, which can be the edges among closer view and foundation pictures. Figure 4.2b is the edge picture of Figure 4.2a, as the square shape among the most likelihood being stamped red. As edge line recognition is concerned, Hough change Sobel, Laplacian, and Canny edge identifiers are utilized to identify moreover flat or vertical edge position inside the picture. In even lines are recognized in the event that one flat procession is steady over the entire video is taken while a determined level line. The image district is able to be situated with a couple of industrious even shape, which may possibly establish two parallel lines of a square shape.

Both vertical and horizontal lines are distinguished, probably with the square shape with two vertical lines and two even lines, which is considered as an image locale. The edges are identified in each key edge, and in this manner, a mean-edge outline is gotten for image restriction. Combination of casing level image results is connected to shape video image results, all the more decisively. After casting a ballot and summation of square shapes recognized from each key casing, the best square shape will be returned as the video [2, 4–6] image locale. In any case, there are a few shortcomings with these current techniques. For a certain something, if each casing in the video is handled, the time utilization is unsuitable. For another, if some key casings are inspected to disentangle the identification, time coherence data will be halfway lost, and the image locale cannot be recognized and found unequivocally. The proposed method in the accompanying area is to dispose of the two folds of weaknesses.

4.3   APPLICATIONS OF ARTIFICIAL INTELLIGENCE OR MACHINE LEARNING IN THE PROPOSED AREA

The proposed three levels of security to check any individual and at the same time protect the biometric template. It is watermarked iris on the face, the verification of the face is detectable, and the iris with watermark is used to authenticate the person and the protection of biometric (face), as well. It is to build the model iris watermark data, i.e., an algorithm based on log on 1D Gabor used. Dlog Gabor is being convolved with the texture of the transformed iris image, and so the model is generated. This model is called iris code. This iris code is in binary form and is unique for each individual. It is now integrated in the image of the face of the same person to protect the face model, as well as the functioning of the [7] multimode offers two levels of security to check any person and at the same time protect the biometric template. It is iris watermarked in the face, the face for verification and diaphragm with water is used to cross to authenticate the person and the protection of the biometric (face).

The proposed algorithm uses three layers of security to maintain the accuracy of the data, privacy, and the protection of personal information. The first layer of security performs technical watermarks to hide personal information of the user; any data that we consider may be a fact of the text (ATM pin, aadhaar cards, date of birth) or can be an image (impression of) of retinal fingers, DNA, and so on. The second layer is applied to encrypt the secret image using encryption and the third layer to

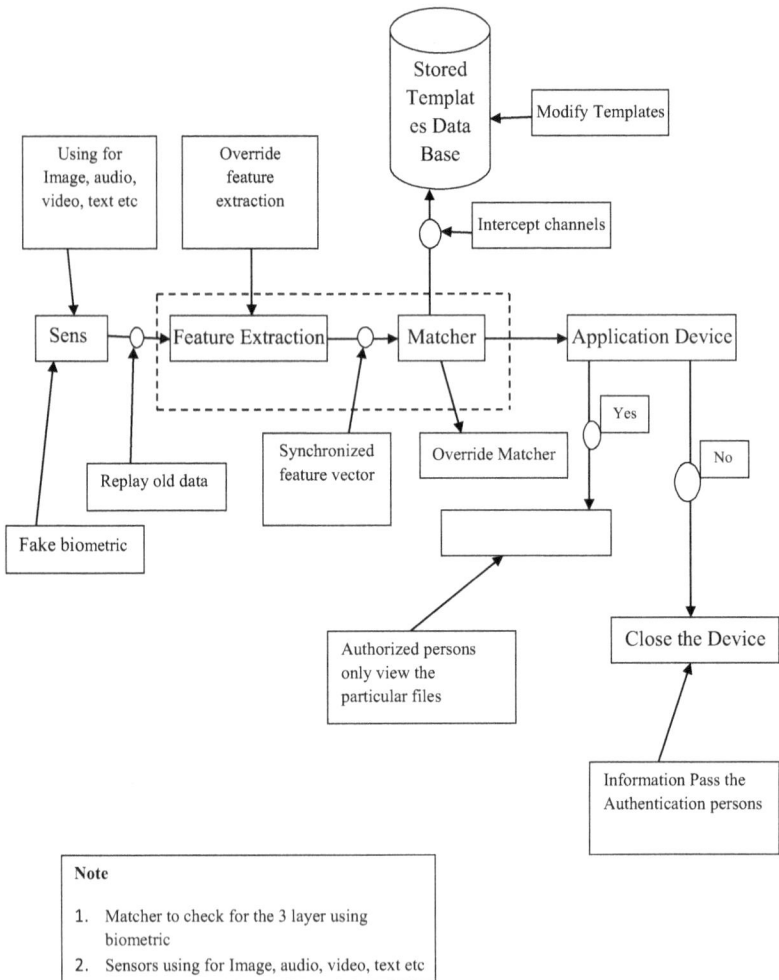

FIGURE 4.7 Proposed work overall architecture.

hide the image with watermark and image secret in the coverage through the image steganography algorithm (Figure 4.7).

4.3.1 PROPOSED ALGORITHM OR METHODOLOGY FOR MAKING MACHINE INTELLIGENT

4.3.2 IMPLEMENTATION RESULTS AND DISCUSSION

4.3.2.1 Steganography Algorithm

To identify concealed messages, an association should effectively screen and organize traffic, which is time-and processor escalated. In any case, the individuals who know about the system's typical traffic examples can just search for changes, for example, expanded development of vast pictures over the system, which may warrant further itemized examination. It is likewise savvy to have – and effectively uphold – a security approach that plainly diagrams adequate use, what information types can and cannot be sent over the system, and how it ought to be ensured. Likewise, confined unapproved programs boycott the utilization of unapproved encryption and [28] steganography in the working environment and consider constraining the measure of postboxes.

At long last, consider deciding if workers who manage secret data ought to approach vast media documents, especially picture, video, or sound records that are to be posted on your Web page. Pernicious gatherings could utilize steganography to pass data by means of such records to an outsider with access to your site. Why not utilize steganography further bolstering your good fortune by utilizing computerized watermarks, a type of steganography, to copyright your Web-open media records? You can even utilize it to conceal framework passwords or keys inside different records to give a progressively secure capacity area.

4.3.2.2 Artificial Neural Network Algorithm

Artificial neural network (ANN) is used for preparing the neural system; there are two gatherings of highlight vectors: one from the first pictures database and the other from the manufactured pictures database. Three hundred component vectors (150 from unique and 150 from replicated pictures database) are utilized to prepare the back

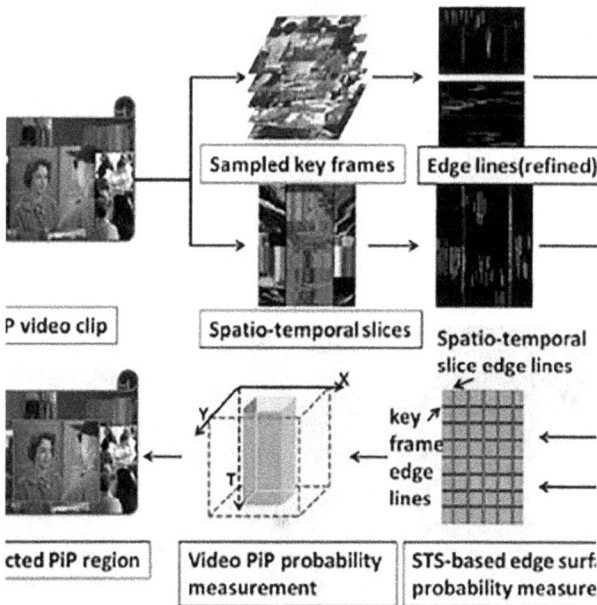

FIGURE 4.8 Image detection for override feature extraction-based edge surface probability measurement.

engendering neural network. The ANN utilized comprises of 30 input neurons and 1 yield neuron with 6 concealed neurons. Staying 300 are utilized to test the ANN.

The video copy detection [9–11] generally focuses on detecting frame level, and image recognition can be representation as an edge image. Figure 4.8 shows sampled key frames and override extracted from the video clip. Vertical and flat edge lines are identified and sophisticated the key frames with the probabilities of four surfaces being measured by the images.

While a video can be investigation as a grouping of pictures, and the image locale through a video cut is actually decided, a period estimation can be acquainted with expanding the model of picture as appeared in Figure 4.8. The central object of video picture location, in this manner, stretches out from recognizing a square shape speak by four edge lines among forefront and foundation pictures to distinguishing solid shapes comprising of four edge surfaces among frontal area and foundation recordings.

While probability estimation of four surfaces is definite, it is dull to [2, 4, 6, 8] remove edges from all edges of the video. Accordingly, certain AI techniques are used to quicken video picture disclosure. In edges of triad, key edges are perceived, and each key edge speaks to one-dimensional line in an edge surface (shown in Figure 4.9). By choosing the probabilities of the lines, the likelihood of the edge surface can be gotten. Notwithstanding, consecutive association of the video cannot be saved by these isolated even lines alone; consequently, it is most essential

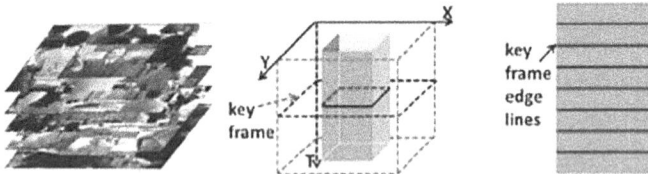

FIGURE 4.9 Key frame-based traditional edge surface probability measurement. (a) Sampled key frames. (b) Key frame for video Pi P. (c) Traditional edge surface probability measurement.

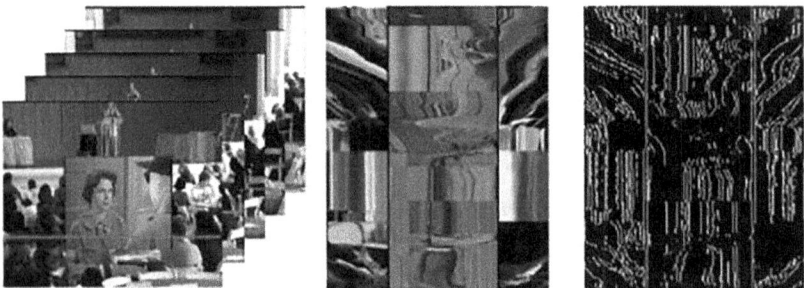

FIGURE 4.10 Synchronized feature vector based on video image detection. (a) A series of video frames. (b) Spatiotemporal slice. (c) Vertical edge image of (b).

to erroneous estimation probability of the edge surface in Figure 4.10 especially in recordings with continuous scene exchanging.

A video can be investigated when there is a succession of pictures through spatial measurement (X,Y) and fleeting measurement T; it can be examined when there are a lot of pictures with measurement (X,T) or (Y,T). For instance, a similar line of an equivalent y (noticeable red in Figure 4.6a–) is selected from all the video outlines, and it is required that a spatio transient cut is correspondingly delivered with measurement (X,T) of the entire video (Figure 4.6b).

Figure 4.11 illustrates the vertical lines that can be found from the spatio-worldly cuts. These sorts of vertical lines, located in the edge surfaces among frontal area and foundation recordings, are utilized at the same time through the level lines as of key casings to decide the likelihood of the edge surface.

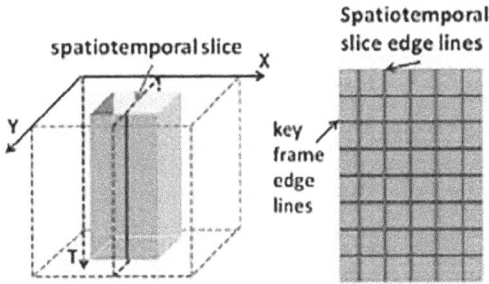

FIGURE 4.11 Sample of spatiotemporal slice. (a) Spatiotemporal slice for video PiP. (b) STS-based edge surface probability measurement.

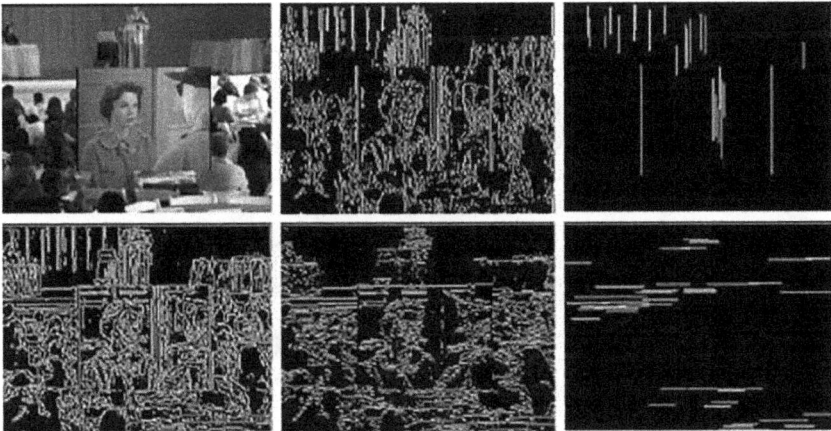

FIGURE 4.12 Raspberry Pi used for video detection edge line extraction. (a) Original image. (b) Canny edges. (c) Vertical canny edges. (d) Horizontal canny edges. (e) Refined vertical edges. (f) Refined horizontal edges.

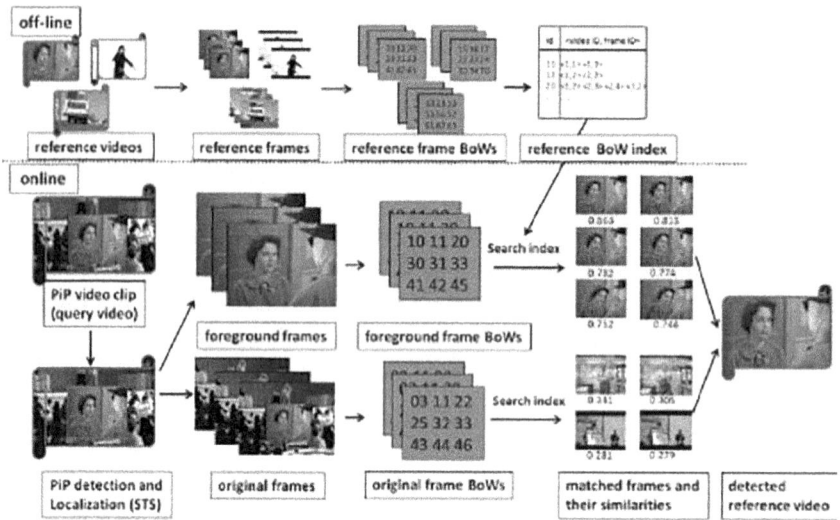

FIGURE 4.13　Raspberry Pi used video copy recognition system.

The vertical and flat edge pictures are extricated separately, which diminishes the figuring of angle bearings by dropping the first eight headings and the present two bearings. The vertical and flat watchful edges are appeared in Figure 4.12c–d individually, among the first vigilant edges appeared in Figure 4.12b. Now a fairly small edge for shrewd edge determination is picked for review rate, raising the likelihood of genuine edges and commotion edges also. At that point, image extraction is intended to join little neighboring edge portions while evacuating short disconnecting ones decreases the uproarious edges with the goal that less edge lines will be picked for further systems.

Figure 4.13 illustrates of the Raspberry Pi utilizing video duplicate recognition framework, which is broadened. The picture is distinguished and limited (whenever identified) in each inquiry video. From that point forward are extricated from closer view casings and unique edges separately. Modified lists are utilized to speed up looking through the most related reference [16] outlines that assess with question outlines. These related reference edges will at last direct to the choice of the inquiry

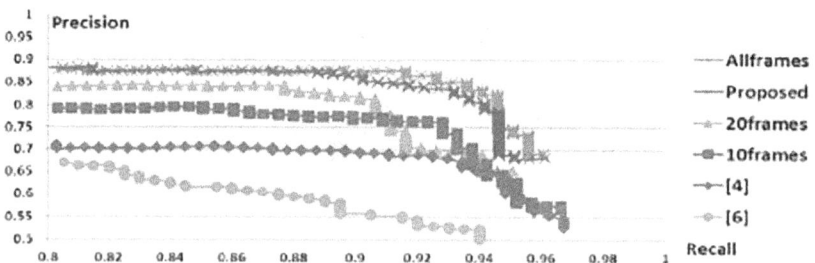

FIGURE 4.14　Illustration of our video copy detection system.

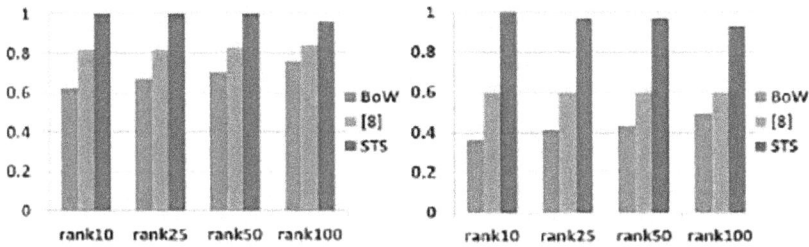

FIGURE 4.15   The retrieval accuracy of video copy detection. (a) Small dataset. (b) Large dataset.

video whether a duplicated video is replicated or reference [17] video is replicated. The execution of the video duplicate recognition framework will be assessed.

The identification accuracy execution with the method [6, 15] of utilizing all edges has the most excellent execution yet with a lot elevated time usage. The projected strategy is shown in Figure 4.14 through right confinement discernible in illumination blue and incorrect restriction observable in red. The reason is that these square shape articles are difficult to be recognized. The review accuracy bends of various strategies are appeared in Figure 4.15, while the handling time can be seen that is assessed with identifying picture locales on edge level; duplicate identification over the entire video has raised exactness and lesser time use. Also, through edge line modification, even raised exactness and lesser time usage is able to be confirmed for video.

4.4   FINDINGS WITH OUTCOMES AND BENEFITS TO THE SOCIETY

IoT utilizing AI is considered as the greatest wilderness, which can improve our lives in numerous viewpoints. Those gadgets that have never been arranged can get associated and react simply like the savvy gadgets do, for example, your vehicle, cooler, and home speakers. IoT and AI are set to change our [13] reality totally. How about we examine the accompanying advantages of IoT according to the business' point of view:

1. Professional resource utilization
2. Shortened human efforts
3. Less the cost and convey productivity
4. Real-time promote
5. Resolution analytics
6. Better customer understanding
7. High-quality records

4.5 CONCLUSION

In this chapter, we have adjusted the IoT and artificial insight AI framework parts to the Raspberry Pi. At that point, following the Raspberry Pi, the vulnerabilities and countermeasures to relieve them were featured. These conceivable outcomes in Refs. [16, 17] are shown for verifying IoT frameworks. We apply this underlying reference model to the revealed IoT security dangers to speak to instantiations of the IoT security hazard idea. At last, we have indicated how the created model could be utilized while developing the genuine word IoT framework so as to give secure support of the clients.

4.6 LIMITATIONS

The present reference display encases couple of confinements. Right off the bat, it essentially covers the framework assets and their vulnerabilities; however, it leaves the examination of business assets (i.e., information traded in the IoT artificial insight AI frameworks, business tasks) and their security criteria away. As to security hazard investigation, we have decided on the vulnerabilities; however, further work is required to feature the profile of the danger delegates and their assault strategies, just as the impact on the IoT framework and business assets. On the framework counter-measure height, we make a supposition that to treat IoT security peril, one goes for broken decrease choice; anyway it is additionally fundamental to comprehend the significance of other taking care of choice (e.g., chance evasion, maintenance, or exchange). At long last, in our proposition, we do not separate between the security necessities and controls.

4.7 INFERENCE

IoT AIML [24, 27] security reference model encourages us to absolutely target coun-termeasure strategies to the IoT framework's layers amid improvement process so as to cover existing vulnerabilities. Such methodology of adjusting IoT artificial knowl-edge AI framework [6] segments to the Raspberry Pi sets gives us chance to fabricate a safe and ensured design for the IoT artificial insight AI framework without any preparation. As it was appeared, if there should arise an occurrence of disregard-ing conceivable security dangers, we put profitable information under risk and gave administration inconsistent and shaky. Thus, such delicate frameworks can get new dangers stack associated with advantageous administrations given by the IoT AIML systems.

4.8 FUTURE WORK

In later research, we intend to reinforce the proposed reference display with the meaning of the express rules for the IoT intelligence ML systems resource, haz-ard, and hazard countermeasure ID, just as the technique for the security exchange of examination. We have likewise intended to put our hands-on case (IoT artificial knowledge AI home app) on increasingly exact infiltration testing, for which extra writing and apparatuses study will be required, as we will attempt to steal information from secured [21] with created structure framework. Ideally, amid such tests, the relationship among the IoT intelligence ML systems, their vulnerabilities, and proposed countermeasures will be investigated in a progressive gritty way and give us more indicating further system advancement.

REFERENCES

1. Mani Malek Esmaeili, Mehrdad Fatourechi, and Rabab Kreidieh Ward. 2011. "A robust and fast video copy detection system using content-based fingerprinting." *IEEE Transactions on Information Forensics and Security* 6, 1, 213–226.
2. P. Karthika and P. Vidhya Saraswathi. 2020. "IoT using Machine Learning Security Enhancement in Video Steganography Allocation for Raspberry Pi." *Journal of Ambient Intelligence and Humanized Computing* 11, 7, 1–15, DOI: 0.1007/s12652-020-02126-4.
3. Jaap Haitsma and Ton Kalke. 2012. "A highly robust audio fingerprinting system." In *Proceedings of the International Symposium on Music Information Retrieval*, 107–115.
4. Menglin Jiang, Yonghong Tian and Tiejun Huang. 2012. "Video copy detection using a soft cascade of multimodal features." In *Proceedings of the IEEE International Conference on Multimedia and Expo (ICME'12)*, 374–379.
5. Ganesh Babu, R. and Amudha, V. 2015. "Analysis of Distributed Coordinated Spectrum Sensing in Cognitive Radio Networks." *International Journal of Applied Engineering Research* 10, 6, 5547–5552.
6. Hong Liu, Hong Lu and Xiangyang Xue. 2013. "A segmentation and graph-based video sequence matching method for video copy detection." *IEEE Transactions on Knowledge and Data Engineering* 25, 8, 1706–1718.
7. Jingkuan Song, Yi Yang, Zi Huang, Heng Tao Shen and Richang Hong. 2013. "Multiple feature hashing for large scale near-duplicate video retrieval." *IEEE Transactions on Multimedia* 15, 8, 1997–2008.
8. Kasim Tasdemir and Enisetin A. Cetin. 2014. "Content-based video copy detection based on motion vectors estimated using a lower frame rate." In: *Proceedings of the Signal, Image and Video Processing*, Springer, Berlin, pp. 1049–1057.
9. P. Karthika and P. Vidhya Saraswathi. 2019. Digital video copy detection using steganography frame based fusion techniques. In: Pandian D., Fernando X., Baig Z., Shi F. (eds) *Proceedings of the International Conference on ISMAC in Computational Vision and Bio-Engineering 2018 (ISMAC-CVB)*. ISMAC 2018. Lecture Notes in Computational Vision and Biomechanics, vol 30. Springer, Cham, https://doi.org/10.1007/978-3-030-00665-5_7.
10. P. Karthika and P. Vidhya Saraswathi. 2017. "A survey of content based video copy detection using big data."*International Journal of Scientific Research in Science and Technology* 3, 5, 114–118.
11. P. Karthika and P. Vidhya Saraswathi. 2017. "Content based video copy detection using frame based fusion technique." *Journal of Advanced Research in Dynamical and Control Systems* 9, SP-17, 885–894.
12. A. Nedumaran and R. Ganesh Babu. 2020. "MANET Security Routing Protocols Based on a Machine Learning Technique (Raspberry PIs)." *Journal of Ambient Intelligence and Humanized Computing* 11, 7, 1–15.
13. Pei-Yu Lin, Bin You and Xiaoyong Lu. 2017. "Video exhibition with adjustable augmented reality system based on temporal psycho-visual modulation." *EURASIP Journal on Image and Video Processing* 2017, 7.

14. Ganesh Babu, R. and Amudha, V. 2018. "Comparative Analysis of Distributive Firefly Optimized Spectrum Sensing Clustering Techniques in Cognitive Radio Networks." *Journal of Advanced Research in Dynamical and Control Systems* 10(9), 1364–1373.

15. Bo-Yi Sung and Chang-Hong Lin. 2017. "A fast 3D scene reconstructing method using continuous video." *EURASIP Journal on Image and Video Processing* 2017, 18. DOI:10.1186/s13640-017–0168-3.

16. Yinghao Cai, Ying Lu, Seon Ho Kim, Luciano Nocera and Cyrus Shahabi Cai. 2017. "Querying geo-tagged videos for vision applications using spatial metadata." *EURASIP Journal on Image and Video Processing* 2017, 19.

17. Nan Nan and Guizhong Liu. 2015. "Video copy detection based on path merging and query content prediction." *IEEE Transactions on Circuits and Systems for Video Technology* 25, 1682–1695.

18. P. Vidhya Saraswathi and M. Venkatesulu. 2013. "A secure image content transmission using discrete chaotic maps." *Jokull Journal* 63, 9, 404–418.

19. Y. Asnath Victy Phamila and R. Amutha. 2014. "Discrete cosine transform based fusion of multi-focus images for visual sensor networks" *Signal Processing* 95,161–170.

20. Om Prakash, Richa Srivastava, Ashish Khare. 2013. "Biorthogonal wavelet transform based image fusion using absolute maximum fusion rule" In *Proceedings of 2013 IEEE Conference on Information and Communication Technologies (ICT)*.

21. K. Sharmila, S. Rajkumar, V. Vijayarajan. 2013. "Hybrid method for multimodality medical image fusion using discrete wavelet transform and entropy concepts with quantitative analysis." In *International Conference on Communication and Signal Processing (ICCSP)*, IEEE, April 3–5, 2013.

22. Lixin Liu, Hongyu Bian and Guofeng Shao. 2013. "An Effective Wavelet-based Scheme for Multi-focus Image Fusion." In *IEEE International Conference on Mechatronics and Automation (ICMA)*.

23. R. Mahmoud, T. Yousuf, F. Aloul and I. Zualkernan. 2015. "Internet of things (IoT) security: Current status, challengesand prospective measures." In *2015 10th International Conference for Internet Technology and Secured Transactions (ICITST)*, 336–341.

24. Michael Negnevitsky. 2011. *Artificial Intelligence: A Guide to Intelligent Systems.* Pearson.

25. I. Kotenko, I. Saenko, F. Skorik and S. Bushuev. 2015. Neural network approach to forecast the state of the internet of things elements. In 2015 *XVIII International Conference on Soft Computing and Measurements (SCM)*, 133–135.

26. Ganesh Babu, R. and Amudha, V. 2014. "Spectrum Sensing Techniques in Cognitive Radio Networks: A Survey". International Journal of Scientific and Engineering Research 5, 4, 23–32.

27. Prabu, S., Ganesh Babu, R., Jewel Sengupta, Rocio Perez de Prado and Parameshachari, B.D. 2020. "A Block Bi-Diagonalization-Based Pre-Coding for Indoor Multiple-Input-Multiple-Output-Visible Light Communication System." Energies, MDPI AG 13(13), 1–16.

28. Adeel Jawed and Atanu Das. 2015. "Security enhancement in audio steganography by RSA algorithm." *International Journal of Electronics & Communication Technology (IJECT)* 6, 139–141.

5 The Implementation of the Concept of Lean Six Sigma Management

Sławomir Świtek and Ludosław Drelichowski
The University of Economy

Zdzisław Polkowski
Jan Wyzykowski University

CONTENTS

5.1   INTRODUCTION

The Industrial Control System Market globally was valued at USD 86.19 billion in 2017 and is expected to reach a value of USD 143.1 billion by 2023, recording a CAGR of 9.1% during the forecast period (2018–2023) [1]. The industrial control systems market protects industries, such as electric, water, transportation, oil and gas, chemical, food and beverage, automotive, and so on.

ERP (enterprise resource planning) is a software-mediated management of core business processes containing all facets of an operation. In order to provide the access to current data, the scheduling system in an enterprise should be integrated with other systems, both in the field of organizational and technical preparation and production control i.e.:

- ERP
- MES (manufacturing execution system)
- SCADA (supervisory control and data acquisition) (see Figure 5.1)

Integration of the systems gives possibility to look into "real-time" data and see actual status of the factory including shop floor level. Key data are transferred back from the production floor to superior systems and converted into critical KPIs (key process indicators) to stipulate appropriate business decisions [2].

ERP uses market-related data (sale forecast, client orders) to create master plan of orders. Functions such as material requirements planning, purchasing, capacity planning, and inventory are needed to prepare the plan. Costs are generated in relation to the planned activities. Product structures and production orders are transferred from ERP system to MES to allow monitoring of production flow, detailed scheduling, and resources utilization. In return, data concerning the state of orders execution and the state of resources are pulled back to ERP system refreshing inputs to capacity utilization and planning.

MES provides "real-time" data while SCADA is used to control assets, which are geographically dispersed in places where centralized data control and acquisition is critical to system operations.

Where does lean six sigma come from? The pure six sigma has its origin at Motorola [3], a little bit later it was adopted by Jack Walch at GE and developed there not as quality initiative, but as an entire business improvement philosophy [4]. Over the years, according to business good practice diffusion, it was applied in many sectors globally, not exclusively in industry. Pepper and Spedding, cited by Switek & Drelichowski [5], attempted to dimension the concept of TQM (total quality management), six sigma, and the hybrid lean six sigma approach extended to lean manufacturing techniques. The basic criteria that distinguish problem-solving concepts are focused on quality, scientific approach to problem-solving, and team approach presented in the following figure in the form of the so-called Joiner's triangle.

The authors pointed out that the TQM concept strongly emphasized teamwork, because all employees work for quality starting from the company's top management; in six sigma, the development of the method of problem-solving occurred, i.e., the DMAIC process [6] (it is basically Deming's PDCA cycle enriched with appropriate tools and improvement methods); and in the lean six sigma concept, the scientific approach has also a central position along with the system approach, although it must be admitted that the development of this methodology takes place in all directions [7]. While six sigma looks for business and economic reasons to push for perfection (zero defects policy), the lean component in six sigma means necessary speed to solve the challenge [8].

In general, the use of ERP/MES/SCADA and HMI systems can support the six sigma methodology in several potential areas.

The added value of the integration of ERP/MES/SCADA with HMI systems is to obtain process data and automatically transfer it between systems in both directions. In addition, the automation and exchange of digital information without human intervention increase the accuracy of the data. Data accuracy is addressed in the measure phase of the six sigma methodology.

Many problems related to production management result from the lack of knowledge about the actual status of orders on the production floor. They usually concern the description of production technology, material availability or completeness of the plan, allocation of resources, and understanding of bottlenecks. Integrated ERP/MES/SCADA systems can enforce the aspects that are placed in the standard operating procedure (SOP), which is the basic document in six sigma used to standardize processes. Standardization reduces volatility and thus the stability of processes that have fewer opportunities to produce defects (reduction of DPMO – defects per million opportunities) [9].

Real-time data obtained via MES (ERP itself is not created for this purpose) confirm the achievement of business goals or show deviations from the goals set. To analyze these deviations, aggregated data from all systems are needed. In order to verify whether the production is carried out as intended, the statistical process control (SPC) technique is used. It is based on the comparison of production parameters with the deviation tolerance for them (a Shewhart chart). In the case where the MES is equipped with a machine operation control module (SCADA), it is possible to configure the system so that it alerts the indicated user or disables the production machines when certain boundary conditions of a given parameter are exceeded. In this way, an out-of-control action plan (OCAP) is called immediately, which reduces losses. The creators of the six sigma concept have already paid particular attention to the fact that the greater the costs associated with the creation of a defective article are, the later the defect is detected. The measures taken to organize the production process should therefore concentrate on identifying any deficiencies as soon as possible. As you can see, ERP/MES/SCADA with HMI (human–machine interface) are helpful to realize it as well. This typically corresponds to the actions taken in the control phase.

HMI may request confirmation of the operator's certification for specific production. In this case, the operator is obliged to confirm his identity by logging into the system. If his/her access is approved, the operator has to go through a series of steps familiarizing with the necessary paperless instructions and making him/her aware of critical quality and safety requirements. If the operator has confirmed his/her knowledge, only the appropriate tools or process equipment is activated. Operator certification or work center certification is an old concept adopted by six sigma into DMAIC's control phase.

The aim of the work is to answer the question included in the title of the article. While implementing lean six sigma, a company often needs to reinvent its approach to collecting data on different levels of organization. Although there are many papers published on lean six sigma and ERP/MES/SCADA separately, there is a dearth of literature connecting these two aspects. As a research method, a comparative analysis was used to draw conclusions.

5.2 COMPARATIVE ANALYSIS OF CHOOSING ENTERPRISE RESOURCE PLANNING SYSTEMS

For the purpose of the study, a report available through Internet website [10] was used. It contains specific data describing 48 different ERP systems commonly used in Poland. In order to make comprehensive comparison of the systems, 580 standard functionalities have been defined, and appropriate YES or NO information has been added to know whether the considered ERP systems contain the functionality or not.

On the website mentioned above, there is also located dictionary where researchers can make searching of terms (ERP functionalities) for his/her further analysis. For standardization, definitions of the terms used are cited from the dictionary in the work. The authors of the article selected words "lean" and "six sigma" to find terms which contain them. In return, four general functions were displayed and summarized as follows:

Quality control in an enterprise/quality management (ISO 9000, GMP).

Quality management in an enterprise is usually used to create a library of procedures, a catalog of activities and operations aimed at maintaining high quality of production and services. It also includes maintaining historical data, issuing necessary documents, compiling quality data statistics, and registering quality maintenance costs. The software should ensure that the ISO 9000 or derivatives are handled as required. In the case of food and pharmaceutical production, it is necessary to support specialized quality assurance procedures, e.g., HCAAP and GMP. Support for quality control can be built into the ERP system or available as an external system. The quality control module can be directly linked to the production and the material flow control system. Quality control techniques are increasingly being used to achieve the best possible quality of products under the slogan "several in a million" defective components. One of them is six sigma.

According to the graph, couple of functions get a quite low representation score of 30%–40%, while for other it is larger by 20%. This concerns basic ISO 9001 standard, but also food industry HCAAP standard and defining quality tests. Only 40% of the systems include functionality for pharmaceutical GMP industry (see Figure 5.1).

CPM (corporate performance management) is a set of management concepts, measures, and information technologies that enable defining, monitoring, and optimizing the implementation of personal and departmental goals in conjunction with the strategy of the entire organization. It is the extension of BI (business intelligence) with one of the management concepts, e.g., BSC (balanced scorecard), VBM (value-based management), or six sigma. They allow the use of advanced planning techniques, results consolidation, and reporting. CPM also allows better preparation of strategic decisions. The following features are typically supported:

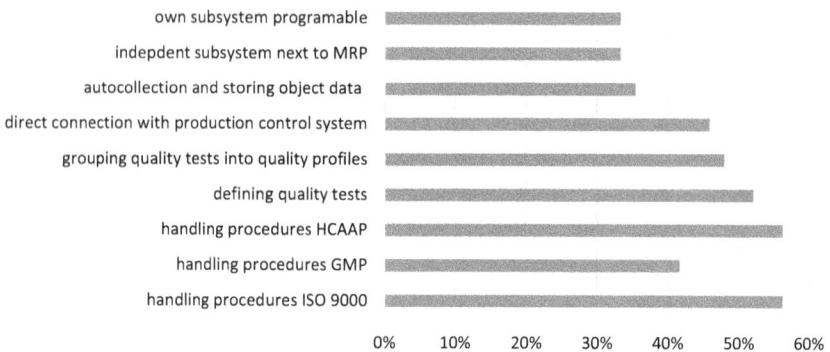

FIGURE 5.1   Quality control functionalities. (Own work.)

- Budgeting, strategic planning, forecasting
- Modeling and optimization of profitability
- Defining and managing score cards
- Financial consolidation
- Financial, tax, and statistical reporting

Strategic planning allows to develop long-term plans and forecasts using statistical techniques and linear programming methods based on forecasts and historical data. This is often made as scenario planning.

BSC is a management concept that describes and specifies what should be measured in order to be able to assess the company's effectiveness in implementing the strategy as well as the adequacy of the strategy itself. The company's goals are presented in the form of a map of their mutual cause-and-effect relationships and measures distinguished in four perspectives: financial, customer, internal processes, and development. The BSC concept is based on the following assumptions:

- Measurement of achievements is a key condition for management efficiency.
- measurement of financial results should be supplemented by measuring the status in other perspectives of the company's operations.
- Goals and measures should be related to activities and budgets, so that they constitute a coherent plan of the company's operations.

Managerial cockpit, knowledge repository, and indicator repository are also considered advanced business decision support functions. Management cockpit is a method of presenting the results of the current enterprise activity in the form of simultaneous presentation of many tables with partial results and indicators. The form of the presentation resembles the placement of indicators in the cockpit of the aircraft.

Knowledge repository is a catalog containing a description of the proposed management techniques, supported by software along with IT tools. The repository of indicators is a list of indicators for company status prepared for immediate use.

Only half of the analyzed ERP systems propose BSC and strategic planning solutions while other functions are much more nested in their standard configurations (see Figure 5.2).

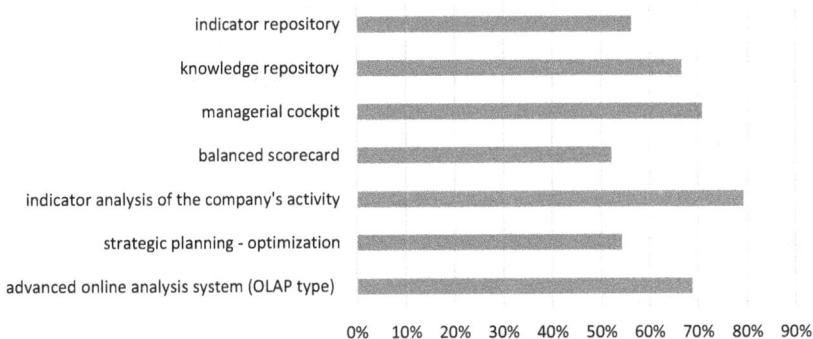

FIGURE 5.2 Advanced business decision support functions. (Own work.)

Production type handled

Not every software package supports all types of production. It is important to carefully match the package to the nature of the production process in individual organizational units. There are several basic types of production:

Design to order (DTO) or project to order (PTO) is a single or small lot production for individual customer orders, where products are often developed for a specific customer. Design and technology are usually developed individually for a specific customer order. Moreover, these projects are implemented on the basis of initial incomplete data, starting from the preparation of the offer and its valuation, through the preparation of the budget and the implementation schedule, ending with the supervision over the implementation and settlement of the contract. The repeatability of orders is small or even none.

Engineering to order (ETO) is a production that needs a new design and technology, and often the purchase of materials. Each customer order results in creation of new material items, material lists, and description of technological operations. The customer usually orders company products using existing variants and options made in series. Make to order (MTO) means manufacturing and assembling a batch of products usually in stock for specific production orders run as needed. Orders can be repeated.

Last but not least, repetitive (mass) production is important. It is the manufacture of one or many products according to stabilized technology on lines and cells organized by an object, i.e., arranged in a line, so as to produce one product or a family of products.

Production type is dependent on the enterprise specifics and can be organized and managed by using the following:

Master production scheduling/materials requirement planning techniques

MRP is defined as material requirements planning based on BOM (bill of materials) structure data, inventory information, open production and purchase orders, order status in progress, and a long-term production plan known as MPS (master production scheduling). The MRP standard is currently considered the main method for determining the effective deadlines for the execution of customer orders and orders including production cycles. It allows to control types of production, quantities, and production dates, as well as control and replenish stocks.

Or JIT technique (just in time) in lean management

In an enterprise managed in accordance with principles of lean management, often long-term and medium-term general planning and material requirements planning, in particular, supplies are carried out according to MRP principles, whereas short-term operational management is carried out according to JIT (just in time) rules.

It can be also done as combination of the aforementioned and TOC (theory of constraints) rules in order to manage properly capacity bottlenecks.

The aforementioned chart proves that systems are still better equipped with tools coming from traditional, push production (70%–85%) while functions that fulfill "pull" concepts are represented moderately less (from 50% to up to almost 70%) (see Figure 5.3).

JIT in production

JIT is a short-term management strategy that was created at Toyota Corporation. It consists in minimizing the use of resources that are necessary to obtain added value of the manufactured products. The resources are time, labor, raw materials and semifinished products, machines, and other work equipment. The assumption is that anything that does not add value is a waste, for example, the cost of capital involved in excessive stocks, storage, excessively extensive quality control, and any excessive transportation. Everything must be available exactly when it is needed for a given position and exactly as much as it is needed. The rule applies: do only what is necessary and start at the last moment to make it on time. This is reflected in logistics – stocks at the supplier or in transit, launching the production of the previous operation at the request of the recipient from the next stage, i.e., so-called "sucking" the necessary components. Classic or electronic Kanban cards are often used to realize JIT concept. Inventories are created and stored when they are absolutely necessary. This is the reciprocal of the principles used in the MRP, where purchase/production requests are planned in advance and forwarding them to realization. In case of MRP, orders are "pushed" for execution, whereas JIT relies on "pulling" orders for execution.

Quality is improved to achieve zero defects, and production cycles are minimized by shortening preparation and completion times, waiting times, and in consequence, batch sizes.

As per Figure 5.4, 50% of ERP systems analyzed use production progress summary to track status of manufacturing. If operations are confirmed by employees through

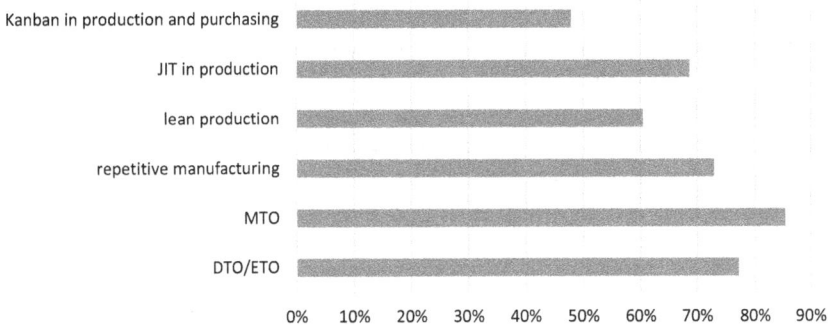

FIGURE 5.3 Production types. (Own work.) DTO, design to order; ETO, engineering to order; JIT, just in time; MTO, make to order.

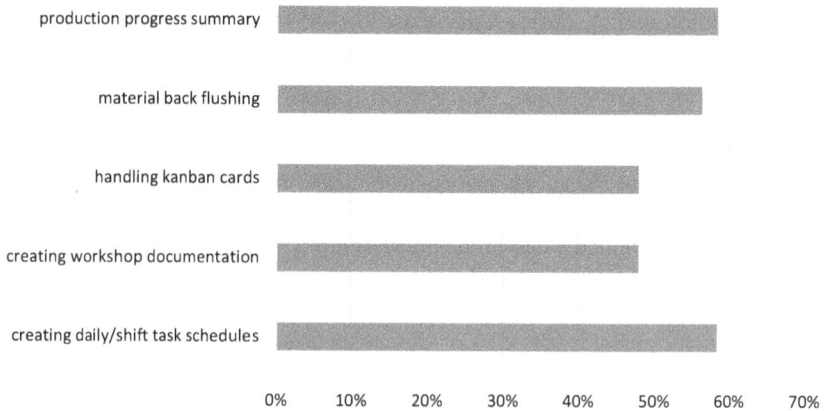

A horizontal bar chart with the following categories (top to bottom) and a horizontal axis ranging from 0% to 70% in 10% increments:

- production progress summary
- material back flushing
- handling kanban cards
- creating workshop documentation
- creating daily/shift task schedules

FIGURE 5.4   JIT supporting functionalities. (Own work.) JIT, just in time.

HMI or by scanning bar codes placed on production orders, then the latest status of the completed activity is known. In order to know the actual progress in "real time," there is MES extension needed. Almost 60% of the ERP solutions make possible to create detailed production schedules typically by machine or work center, where employees can be selected to make specific operation.

5.3   CONCLUSIONS

On the business management level, integration of ERP, MES, and SCADA gives a huge advantage to ERP making "real-time" data accessible. In six sigma world, it means opportunity to make instantaneous decisions to reduce waste based on alarm signals generated, getting real evidence about the process stability or improvement or just getting reliable baseline data for DMAIC projects.

Although it is not a primary goal to ERP/MES SCADA systems to include lean six sigma tools, some of them are included into system functional modules as a kind of extension of already existing basic functionalities, e.g., material management enriched by JIT or Kanban concepts.

The comparative analysis depicted that at least four basic functionalities are repeatedly used by 50%–60% of the analyzed ERP systems – quality control and quality management, corporate performance management, which includes such tools as BSC and BI supporting strong decision-taking process, production type handled (including lean production), and production JIT. What is interesting is that these features are clustered into some ERP systems, whereas others do not propose them at all.

In order to get harmonization of management concepts within an organization, it is recommended to lean six sigma implementers to consider ERP systems familiarized with lean six sigma concepts. Proper IT infrastructure can definitely support lean six sigma, but on the other hand, it cannot be seen as an exhaustive answer to six sigma prerequisites in the implementation phase [11].

REFERENCES

1. Global Industrial Control Systems Market Size, Share - by Control System (SCADA, DCS, PLC, MES, PLM, ERP, HMI), End-user, and Region - Growth, Trends and Forecast (2018–2023), https://www.researchandmarkets.com/research/mfk3v2/global_industrial?w=4, download date February 16, 2019.
2. Kalinowski K., Grabowik C., Kempa W., Paprocka I.: The role of the production scheduling system in rescheduling, Modern Technologies in Industrial Engineering (ModTech2015), *IOP Conf. Series: Materials Science and Engineering* 95 (2015) 012140.
3. Pande P., Neuman R., Cavanagh R.: *The six sigma way: How GE, Motorola and other top companies are honing their performance*, The McGraw-Hill Companies, Inc., 2000, p. 6.
4. Eckes G.: *The six sigma revolution: How General Electric and others turned process into profits*, John Wiley & Sons, 2007, p. 32.
5. Switek S., Drelichowski L.: The Lean startup – a new learning method for organizations? *Studies & Proceedings of Polish Association for Knowledge Management* 89 (2018) pp. 20–32.
6. Eckes G.: *Making six sigma last: Managing the balance between cultural and technical change*, John Wiley & Sons, 2001, p. 12.
7. George M. L.: *Lean six sigma: Combining six sigma quality with lean speed*, McGraw-Hill, 2002.
8. Paris J.: *State of readiness – Operational excellence as precursor to becoming a high performance organization*, GreenLeaf Book Group Press, 2017, p. 32.
9. Liker J., Meier D.: *The Toyota way fieldbook*, McGraw-Hill, 2011, pp. 151–189.
10. https://www.raport-erp.pl, download date February 16, 2019.
11. Gardner K.: *Successfully implementing lean six sigma – The lean six sigma deployment roadmap*, Pinnacle Press, 2014, Chapter 10 – Create supporting infrastructure, pp. 125–135.

6 Design and Performance Analysis of Carrier Depletion PIN Phase Shifter for 50 Gbps Operation

R. G. Jesuwanth Sugesh and A. Sivasubramanian
Vellore Institute of Technology

CONTENTS

6.1 INTRODUCTION

Silicon optical modulators form a vital part of the optical data communication applications. Silicon photonics utilizes the advantages of the CMOS fabrication industry to produce optical components at low cost. Carrier depletion–based modulators are easy to fabricate and promote high-speed operation at low voltages [1]. Mach-Zehnder modulator (MZM) is preferred among others as it is easy to fabricate. It comprises of the phase shifter where the optical signal phase is altered by varying the refractive index by the application of voltage. Thus the phase shifter dictates the performance of the modulator. The core metrics of a phase shifter are capacitance per unit length, efficiency in phase shift, and loss and speed of operation. The voltage required to produce π phase shift in MZM is referred as V_π, and the length at which it is obtained is termed as L_π. $V_\pi L_\pi$ is expected to be minimum such that with shorter length of the phase shifter and at low voltage, π phase shift is obtained [2]. When the length of the phase shifter is small and low voltage is applied, extinction ratio (ER) is reduced. In order to have high ER with smaller phase shifter length and at low voltage, the carrier concentration should be increased. This increases the cost of the modulator and also the optical loss due to free carrier absorption (α).

$\alpha V_\pi L_\pi$ will be higher for highly doped phase shifters that are designed for low $V_\pi L_\pi$ [3]. Waveguide geometry also plays a major role in defining the performance of a phase shifter. With a well-confined waveguide, the loss can be reduced, but at the expense of the efficiency of the phase shifter. Thick slabs in the waveguide promotes high-speed operation but increases the waveguide bend loss. The positioning of the doped regions influences the optical loss and the speed of the operation. Highly doped regions near to the waveguide rib increase the contact of the dopants with the light, thus reducing the V_π and increasing the speed of operation, but at the cost of optical loss. High carrier concentration and ohmic contacts with the aluminum electrodes are formed, reducing the resistance to the PN junction and increasing the capacitance of the junction. With low carrier concentration, the loss and the capacitance are reduced but at the expense of $V_\pi L_\pi$ and speed of operation. Thus, these trade-off conditions have to be considered while designing a phase shifter for the modulator. Various doping patterns, i.e., lateral, vertical, and interdigitated patterns, and concentrations have been experimented by various researchers [4–6]. Lateral PN junction structure is preferred as it is easy for fabrication. PIN junction is favored over PN junction as the inclusion of the intrinsic region reduces the optical loss. For commercial applications, the phase shifter should be able to obtain high ER with low loss and have low $V_\pi L_\pi$, operating at high speed. Simulation analysis is cost-efficient and fast method of designing a phase shifter before hardware prototyping. In this proposed design, the intrinsic gap is kept at 300 nm to reduce the optical loss by reducing the exposure to dopants. Section 6.2 explains the design structure. The carrier concentration is varied from e + 16 to e + 18 cm^{-3}, and the results are analyzed in Section 6.3. This chapter concludes in Section 6.4.

6.2 DESIGN STRUCTURE

Modulators work on single mode especially TE$_1$ and hence the width of 500 nm is selected. The 220 nm (T_{Rib}) thickness provides better optical confinement; at the same time, the rib structure with slab height 90 nm is well suited for phase shifter operation. The light transmission through the waveguide is shown in Figure 6.1. The P++ and N++ wells comprising of higher concentration are situated far from the center to avoid free carrier absorption loss with the concentration of 1e+19 cm^{-3}. The length of the phase shifter is set to 5 mm (L).

The cross-sectional view of the proposed PIN phase shifter is shown in Figure 6.2. The PN region is separated by the intrinsic gap (W_i) of 300 nm not only to reduce the exposure of the dopants to the light but also to be used for the phase shifting operation. In order to maintain the trade-off condition among the parameters but also to obtain high modulation efficiency of the phase shifter, the concentration is varied (e + 16 to e+ 18 cm^{-3}), and the obtained results are tabulated.

6.3 RESULTS AND DISCUSSION

Lumerical software is used for simulation analysis. The electrical circuit for the proposed phase shifter is given in Figure 6.3. The PN regions act as plates of a capacitor when the phase shifter is in reverse bias condition.

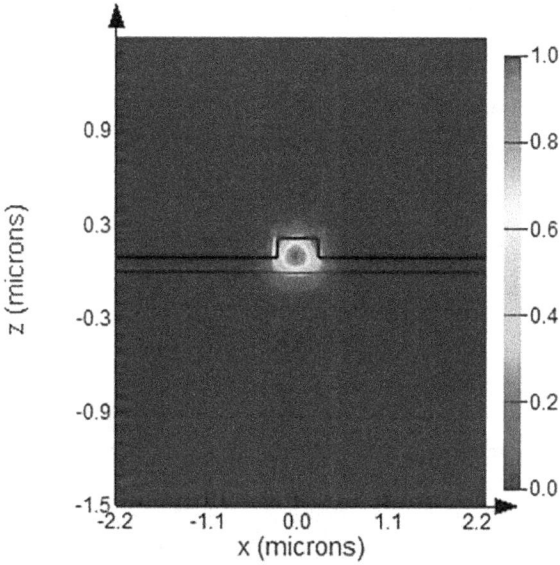

FIGURE 6.1 TE$_1$ mode photonic transmission through the designed waveguide.

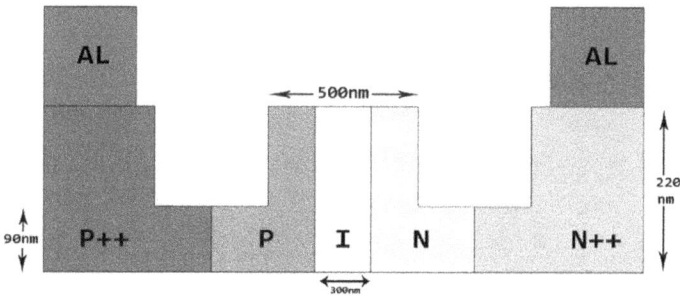

FIGURE 6.2 Proposed PIN phase shifter – cross-sectional view.

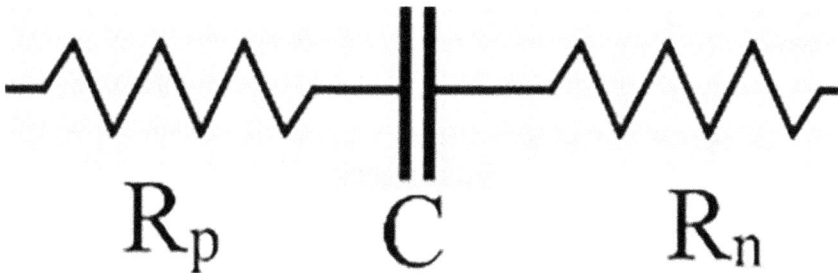

FIGURE 6.3 Simplified electrical circuit diagram for the proposed PIN phase shifter.

With the change in the number of carriers based on the carrier concentration, the capacitance of the proposed phase shifter varies, and it is calculated based on equation (6.1):

Capacitance of the junction [7]:

$$C = T_{\text{Rib}} \times \sqrt{(q\varepsilon_o\varepsilon_s - (2(N_N^{-1} + N_P^{-1}) \times (V_i - V))}$$ (6.1)

where

ε_s = relative permittivity,
N_N and N_P = doping densities,
V_i = diffusion potential,
V = voltage applied.

When the carrier concentration is increased, the number of charge carriers in the doped region increases, which in turn increase the capacitance of the region as in Figure 6.4. When the carriers are increased, less driving voltage (V) is needed to produce the effective index change Δn_{eff} (V) required for the phase shifting operation for the wavelength (λ). The phase shift obtained for the corresponding voltage is calculated as in equation (6.2):

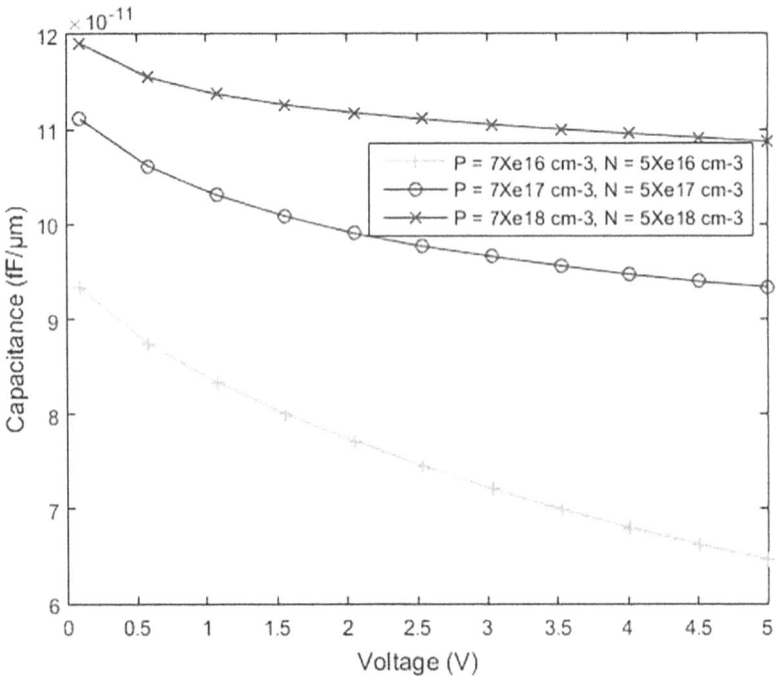

FIGURE 6.4 Capacitance versus voltage for various carrier concentrations in the proposed phase shifter.

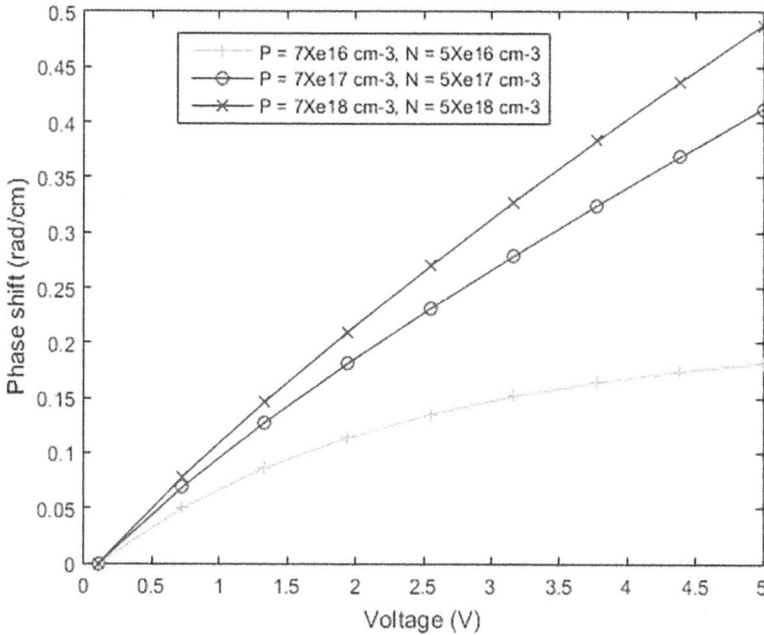

FIGURE 6.5　Phase shift vs voltage.

$$\varphi = \frac{2\pi \Delta n_e ff(V)}{\lambda L} \tag{6.2}$$

Figure 6.5 shows the phase shift variation for the corresponding voltage. With the increase in voltage applied, the carriers get depleted from the center, which causes an increase in effective index change and leads to increase in phase shift. As the dopant exposed to the light propagation is reduced by the application of the voltage, the loss of the phase shifter is reduced (Figure 6.6). From Figure 6.6, it is also inferred that the loss increases with the increase in carrier concentration due to free carrier absorption loss effect.

For further analysis, an unbalanced Mach-Zehnder modulator with the arm length variation of 100 μm and the proposed PIN phase shifter on a single arm was used. The transmission wavelength =1553.5 nm was used for the detailed study.

The analysis was performed using a pseudo random sequence with NRZ coding technique at 20 Gbps bit rate. The ERs and BER obtained for the respective voltages for the carrier concentrations in the proposed PIN phase shifter are tabulated in Table 6.1.

From Table 6.1, it is inferred that the $P = 7e + 17$ and $N = 5e + 17 \, cm^{-3}$ provide higher ER and acceptable BER at lower $V_\pi L_\pi$. When the carrier concentration was increased to $e + 18 \, cm^{-3}$, the ER increased, but due to increase in optical loss, the BER also increased which was not under acceptable communication standards.

A proper eye in an eye diagram ensures that the device can be utilized for a high data rate transmission. The eye crossing in the eye was calculated to be above 20%,

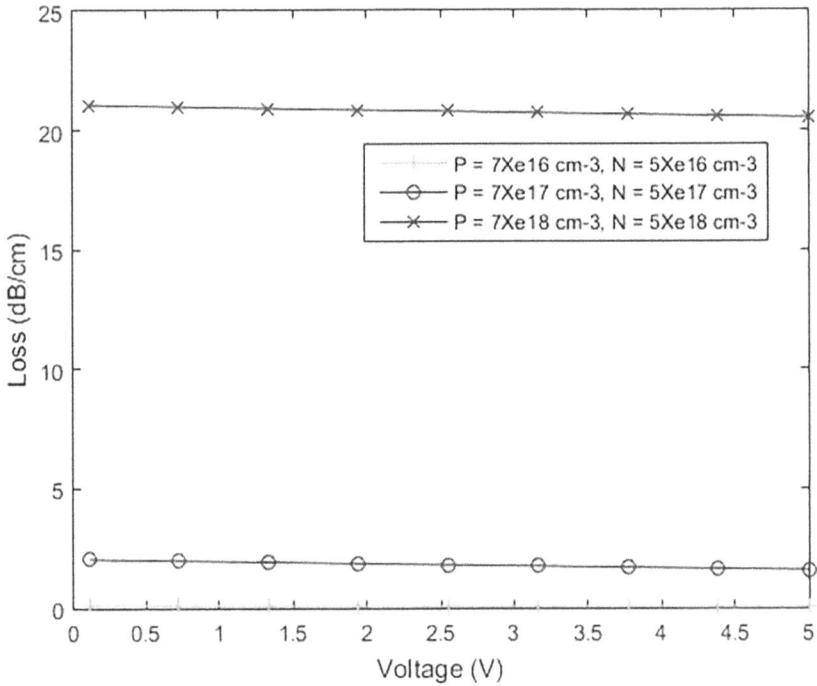

FIGURE 6.6   Loss versus voltage.

TABLE 6.1
Carrier Concentration Variation Results

P (cm^{-3})	N (cm^{-3})	Voltage (V)	ER (dB)	BER
7e + 18	5e + 18	0.5	7.894	0.029
		1	9.7709	0.0001
		2	14.5718	0.009
		3	17.5774	0.0011
7e + 17	5e + 17	0.5	2.6351	0.0002
		1	5.599	2.67e^{-08}
		2	**16.1772**	**8.33e^{-15}**
		3	13.3283	4.36e^{-14}
7e + 16	5e + 16	0.5	1.7817	0.0007
		1	3.1032	1.22e^{-06}
		2	5.3936	4.53e^{-11}
		3	7.3114	5.83e^{-14}

BER, bit error rate; ER, extinction ratio.
The best results obtained were bolded.

and with lower jitter, the proposed phase shifter can be used for high-speed transmission of data. From the eye, it is inferred that the timing for "one" signal pulse is lower than that of "zero" signal pulse; hence, duty cycle distortion (DCD) was calculated as given in equation (6.3):

$$DCD = \frac{100 \times (t_r \ \& \ t_f \ difference \ at \ the \ eye \ center)}{Bit \ period} \quad (6.3)$$

DCD defines the time distortions between the signal zero and one. Error in receiving the bits occurs when the DCD value deviates from the ideal 0%. With the widened and open eye in Figure 6.7, the DCD value was found to be 3.17%. This makes the proposed phase shifter feasible for high-speed data transmission. The bit rate was increased to 50 Gbps, and the results are tabulated in Table 6.2. When the bit rate is increased, the error in differentiating the bits increases, which leads to the increase in BER. The data rate was restricted to 50 Gbps taking into consideration of the communication standards.

The obtained results are compared with various published articles in Table 6.3. The rib region was undoped [4] to minimize the loss, but as the doping region in contact with the optical region is reduced, higher voltage (6 V) was required. This increased the $V_\pi L_\pi$ 7.2 Vcm. A vertical PN junction was designed in [8] in order to reduce the modulation voltage necessary for V_π in a 3-mm phase shifter. As electrons have higher absorption in comparison with holes, the P-type region was proposed

FIGURE 6.7 Eye diagram obtained for 50 Gbps ($P = 7e + 17\,cm^{-3}$, $N = 5e + 17\,cm^{-3}$ at 2 V).

TABLE 6.2

ER and BER for Higher Bit Rates ($P = 7e + 17$ cm^{-3}, $N = 5e + 17$ cm^{-3} at 2 V)

Gbps	ER (dB)	BER
20	16.1772	$8.33e^{-15}$
30	16.1352	$4.75e^{-11}$
40	16.1002	$4.13e^{-09}$
50	16.0698	$6.50e^{-08}$

BER, bit error rate; ER, extinction ratio.

TABLE 6.3

Parameter Comparison

Ref	λ (nm)	L (mm)	V	Gbps	α (dB/cm)	$V_\pi L_\pi$ (Vcm)	$\alpha V_\pi L_\pi$ (VdB)
[4]	1549.4	6	6 on each arm	10	–	7.2	–
[8]	–	3	3	10	5	1.08	5.4
[9]	1550	4	3.6	–	5.2 dB	1.44	7.488
[10]	1550	3	~2.7	28	9.1	0.8	12.1
[11]	~1545	1.5	1	50	–	2.4	–
[12]	~1550	5	5.1	56 & 100	–	2.5	–
This work	1553.5	5	2	50	1.8703	1	1.87

to be larger than the N-type region in [9]. Higher doping concentration and vertical slab structure were utilized in [10] to obtain a lower $V_\pi L_\pi$ at the expense of loss. Interleaved PN junction was proposed in [11] to produce many PN junctions to obtain π phase shift at reduced length of 1.5 mm at the expense of ER (5dB). In [12], lithium niobate MZM modulator was designed with 5 mm modulator length which performed at 56 and 100 Gbps to obtain ER of 11 and 5 dB with BER of ~e^{-6} and ~e^{-3}, respectively. The insertion loss obtained was 2.5 dB. The $V_\pi L_\pi$ was twice than the proposed design. From Table 6.3, it is inferred that $V_\pi L_\pi$ is lower for the proposed model and loss is less, which makes it feasible for high-speed data rate commercial applications.

6.4   CONCLUSION

The study provides a detailed overview of selecting the optimum carrier concentration for the proposed design in obtaining high modulation efficiency. Speed of operation, capacitance, loss, and phase efficiency are the metrics taken into account in determining the carrier concentration for the design. Obtaining high ER with decent BER at low $V_\pi L_\pi$ was kept as the objective. Simulations were performed by varying the dopant concentration (P and N) between e + 16 and e + 18 cm^{-3}. The analysis shows that 16 dB ER with 1.8703 dB/cm loss is obtained at 1 Vcm $V_\pi L_\pi$ for 50 Gbps operation, when the carrier concentrations are $P = 7e + 17$ cm^{-3} and $N = 5e + 17$ cm^{-3} and intrinsic gap is 300 nm.

ACKNOWLEDGMENT

Thanks to VIT University (Chennai campus) for providing the resources and support for this work.

REFERENCES

1. Reed, G. T., Mashanovich, G. Z., Gardes, F. Y., Nedeljkovic, M., Hu, Y., Thomson, D. J., Li, K., Wilson, P. R., Chen, S. W., Hsu, S. S.: Recent breakthroughs in carrier depletion based Silicon optical modulators, *Nanophotonics* 3(4–5), 229–245 (2014).

2. Rasigade, G., Marris-Morini, D., Vivien, L., Cassan, E.: Performance evolutions of carrier depletion silicon optical modulators: from PN to PIPIN diodes, *IEEE Journal of Selected Topics in Quantum Electronics* 16(1), 179–184 (2010).

3. Kim, Y., Takenaka, M., Takagi, S.: Numerical analysis of carrier-depletion strained SiGe optical modulators with vertical PN junction, *IEEE Journal of Quantum Electronics*, 51(4), 1–7 (2015).

4. Goi, K., Ogawa, K., Tan, Y.T., Dixit, V., Lim, S.T., Png, C.E., Liow, T.Y., Tu, X., Lo, G.Q., Kwong, D.L.: Silicon Mach-Zehnder modulator using low-loss phase shifter with bottom PN junction formed by restricted-depth doping, *IEICE Electronics Express*, 10, 20130552 (2013).

5. Xu, H., Li, X., Xiao, X., Li, Z., Yu, Y., Yu, J.: Demonstration and characterization of high-speed Silicon depletion-mode Mach-Zehnder modulators, *Journal of Selected Topics in Quantum Electronics*, 20(4), 3400110 (2014).

6. Azadeh, S.S., Merget, F., Romero-García, S., Moscoso-Mártir, A., von den Driesch, N., Müller, J., Mantl, S., Buca, D. and Witzens, J.: Low V π Silicon photonics modulators with highly linear epitaxially grown phase shifters, *Optic Express*, 23(18), 23526–23550 (2015).

7. Chrostowski, L., Hochberg, M. *Silicon photonics design: from devices to systems*, Cambridge University Press, pp. 217–223 (2016).

8. Ogawa, K., Goi, K., Ishikura, N., Ishihara, H., Sakamoto, S., Liow, T., Tu, X., Lo, G., Kwong, D., Lim, S.T., Sun, M.J., Png, C.E.: Silicon-based phase shifters for high figure of merit in optical modulation, *Proc. SPIE 9752, Silicon Photonics XI*, 975202 (2016).

9. Png, C.E., Sun, M.J., Lim, S.T., Ang, T.Y., Ogawa, K.: Numerical modeling and analysis for high-efficiency carrier-depletion silicon rib-waveguide phase shifters, *IEEE Journal of Selected Topics in Quantum Electronics*, 22(6), 99–106 (2016).

10. Maegami, Y., Cong, G., Ohno, M., Okano, M., Itoh, K., Nishiyama, N., Arai, S., Yamada, K.: High-efficiency strip-loaded waveguide based silicon Mach-Zehnder modulator with vertical pn junction phase shifter, *Optics Express*, 25(25), 31407–31416 (2017).

11. Hu, Z., Jin, L., Guo, J. Feng, J.: High bandwidth silicon Mach-Zehnder modulator based on interleaved PN junctions, *Asia Communications and Photonics Conference (ACP)* (2018).

12. He, M., Xu, M., Ren, Y., Jian, J., Ruan, Z., Xu, Y., Gao, S., Sun, S., Wen, X., Zhou, L. and Liu, L.: High-performance hybrid silicon and lithium niobate Mach–Zehnder modulators for 100 Gbit s− 1 and beyond, *Nature Photonics*, 13(5), 359 (2019).

7 Soft Computing Techniques for Blood Bank Inventory Model for Decaying Items with Storage Using Particle Swarm Optimization

Ajay Singh Yadav
SRM Institute of Science & Technology

Anupam Swami
Government Post Graduate College

Navin Ahlawat, Dhowmya Bhatt, and Tripti Pandey
SRM Institute of Science & Technology

CONTENTS

7.1 INTRODUCTION

In the present market, the explosion of choice due to fierce competition means that no company can resist inventory because there are a variety of alternative products for added features. In addition, there is no cut-and-dried recipe, with which the requirement can be determined exactly. Therefore, when an enterprise needs inventory, it must be protected so that the physical characteristics of the inventory elements can be preserved and protected. How an organization stores its stock depends critically on its ability to achieve the ultimate goal of inventory management. An organization collects the same stock in different ways, even generating profits and losses, to achieve different results. Material storage eventually becomes the core of the overall inventory management practice. There are blood storage container stores for the storage and storing of goods and the provision of other related services to encourage distributors and/or manufacturers to maintain the products in a scientific and orderly manner, so that they can achieve their original value, quality, and utility. Keep it intact. It is an integral part of an industrial unit. It serves as the custodian of all materials required by the industrial unit and supplier materials. Sometimes the total requirement of an item is such that the supplier is more likely to buy it than it is stored in its blood stock. If you have purchased at least a certain amount, you may be impressed by the low price offered for the stock. Or you can expect a strong sales season and you want to be prepared in advance so that you do not lose the chance to make big profits. Another may be an imminent strike, subcontracting, or lockout, which may be a threat to a recovery period. He really wants to make up for this period by buying more than he can store in his blood store. In busy markets such as supermarkets, community markets, and so on, storage space for goods is limited. If a good-looking reduction is obtainable for size purchase, or if the purchase cost of goods is higher than other costs related to inventory or demand for very large items, or if there are problems with frequent purchases, then management manages to buy many items and decides to save time. These items cannot be stored in an existing warehouse in the current market. In this case, an additional storage of blood is hired on a rental basis for the storage of surplus goods.

Blood storage is the storage of controlled blood or hazardous substances in blood storage facilities, blood storage cabinets, or similar devices. Improper storage of blood can compromise workplace safety, including heat, fire, explosions, and leakage of toxic gases. Blood storage cabinets are typically used to safely store small amounts of blood in the workplace or laboratory for regular use. These cabinets are usually made of antifungal materials, which accumulate in them and are sometimes in a pack tray to recover dropped material. Blood stores are warehouses commonly used by blood or pharmaceutical companies for the storage of bulk blood. In the United States, the storage and handling of potentially hazardous products must be disclosed in accordance with potential Occupational Safety and Health Administration (OSHA) legislation. Blood storage devices are typically found in workplaces that require the use of nonhazardous and/or dangerous blood. Proper storage is necessary for the safety and accessibility of laboratory technicians.

The particle swarm optimization (PSO) algorithm is based on the social behavior of birds. This algorithm first creates a random population. Each person called a particle is given a speed and a small social network. For all particles, the values of the fitness or objective function are evaluated. There is no cross/mutation based on physical conditions with respect to PSO, but individual optimum for each individual, total optimum in the total population, and neighborhood optimum found by each individual's neighbors are stored for speed and position updates. This process is repeated until maximum generations or convergence is reached.

7.2 LITERATURE REVIEW

Yadav and Swami [1, 2] presented an integrated supply chain model for the degradation of basic products with an adapted linear demand and in a climate of disruption and inflation and a constraint varying in time for a model and the portion size of the female stock. Yadav et al. [3–6] introduced a supply chain warehouse for the expiration of two stocks and inflation and proposals for an inventory model for the deterioration of two stocks and goods with varying costs and deterioration and discussed the psychoanalysis of green supply chain inventory management for warehouse storage and ecological teamwork using a genetic algorithm and sustainability performance using a genetic algorithm. Yadav and Kumar [7] demonstrated the management of the supply chain of electronic components for storage in collaboration with the environment and neural networks. Yadav et al. [8–10] examined the effect of inflation on a two-stock commodity stock which was exacerbated by changing needs and shortages and discussed a model of inventory which was inflationary for the deterioration of goods in two-inventory systems and proposed an obscure store nonmerchandise model before temporarily deteriorating the goods with a conditional late payment permit. Yadav [11] analyzed supply chain management in optimizing warehouses with logistics using the genetic algorithm. Yadav et al. [12, 13] explained the inventory model for two bearings with optimized soft IT functionality. Yadav explained the modeling and analysis of the supply chain inventory model with two-stage economic transfer problems using the genetic algorithm.

7.3 ASSUMPTIONS AND NOTATIONS

In developing the mathematical model of the inventory system, the following assumptions are being made:

1. The demand rate $D(t) = (u_0 + \omega - 1)e^{-(\lambda_0 + \omega_1 - 1)\{t - (t - t_1)H(t - t_1)\}}$, $(u_0 + \omega - 1) > 0$, $(\lambda_0 + \omega_1 - 1) > 0$.
2. The backlogging rate is $\exp[-(\delta_0 + \omega_2 - 1)t]$.
3. The variable rate of deterioration in blood bank storage is taken as $(\theta + \omega_2 - 1)(t) =$, where $0 < (\theta + \omega_2 - 1) \ll 1$.

In addition, the following notations are used throughout this chapter:

$\Upsilon_{BBS}(t) \le$	The inventory level in blood bank storage at any time t
T	Planning horizon
$(r_0 + \omega_3 - 1)$	Inflation rate
ζ_{BBHC}	Holding cost
ζ_{BBDC}	Deterioration cost
ζ_{BBSC}	Shortage cost
ζ_{BBLS}	Opportunity cost
ζ_{BBOC}	The replenishment cost
$BBSTC(t_1, T)$	Blood bank storage total cost

7.4   FORMULATION AND SOLUTION OF THE MODEL

$$\frac{d\Upsilon_{BBS}(t)}{dt} + (\theta + \omega_2 - 1)(t)\Upsilon_{BBS}(t) = -(u_0 + \omega - 1)e^{-(\lambda_0 + \omega_1 - 1)t}, \qquad (7.1)$$

boundary condition $\Upsilon_{BBS}(0) = 0$,

$$\Upsilon_{BBS}(t) = (u_0 + \omega - 1)\left\{\begin{array}{l}(t_1 - t) - \dfrac{(\lambda_0 + \omega_1 - 1)}{2}\left(t_1^2 - t^2\right)\\[2mm] + \dfrac{(\theta + \omega_2 - 1)}{6}\left(t_2^3 - t^3\right)\end{array}\right\}e^{-(\theta + \omega_2 - 1)t^2/2}, \qquad (7.2)$$

The total average cost consists of subsequent rudiments:

1. **Ordering cost in blood bank storage:**

$$BBS_{OC} = \zeta_{BBOC} \qquad (7.3)$$

2. **Holding cost in blood bank storage:**

$$BBS_{HC} = \zeta_{BBHC}\left[\int_0^{t_1}\Upsilon_{BBS}(t)e^{-(r_0+1)t}\,dt\right]$$

$$= (\zeta_{BBHC})(u_0 + \omega - 1)\left[\begin{array}{l}\dfrac{t_1^2}{2} - \dfrac{(3(\lambda_0 + \omega_1 - 1) + (r_0 + 1))}{6}t_1^3\\[2mm] + \left(\dfrac{(\theta + \omega_2 - 1)}{12} + \dfrac{(\lambda_0 + \omega_1 - 1)(r_0 + \omega_3 - 1)}{8}\right)t_1^4\\[2mm] - \left(\dfrac{(r_0 + \omega_3 - 1)(\theta + \omega_2 - 1)}{20} - \dfrac{(\lambda_0 + \omega_1 - 1)(\theta + \omega_2 - 1)}{30}\right)t_1^5\end{array}\right]$$

$$(7.4)$$

3. **Cost of deteriorated units per cycle in blood bank storage:**

$$BBS_{DC} = \zeta_{BBDC}\left[\begin{array}{l}\int_0^{t_1}(\theta + \omega_2 - 1)t\Upsilon_{BBS}(t)e^{-(r_0+\omega_3-1)t}\,dt\\[2mm] + \int_\mu^{t_2}(\theta + \omega_2 - 1)t\Upsilon_{BBS}(t)e^{-(r_0+\omega_3-1)(t+t_1)}\,dt\end{array}\right]$$

$$
= \zeta_{BBDC}(\theta + \omega_2 - 1)
\left[
\begin{array}{l}
\left[
\left(\dfrac{t_1^2}{2} - \dfrac{(r_0 + \omega_3 - 1)t_1^3}{3} - \dfrac{(\theta + \omega_2 - 1)t_1^4}{8}\right) + (u_0 + \omega - 1)e^{-t_1(\lambda + r)}
\right] \\[2em]
\left[
\begin{array}{l}
\left(\dfrac{t_2^3}{6} - \dfrac{(r_0 + \omega_3 - 1)t_2^4}{12} + \dfrac{(\theta + \omega_2 - 1)t_2^5}{40}\right. \\[1.5em]
\left. - \dfrac{(r_0 + \omega_3 - 1)(\theta + \omega_2 - 1)t_2^6}{36} - \dfrac{t_1^2}{6}(3t_2 - 2t_1)\right. \\[1.5em]
\left. - \dfrac{(\theta + \omega_2 - 1)t_1^2}{60}(5t_2^3 - 2t_1^3) - \dfrac{(r_0 + \omega_3 - 1)t_1^3}{12}(4t_2 - 3t_1)\right. \\[1.5em]
\left. - \dfrac{(r_0 + \omega_3 - 1)(\theta + \omega_2 - 1)t_1^3}{36}(2t_2^3 - t_1^3)\right. \\[1.5em]
\left. - \dfrac{(\theta + \omega_2 - 1)\mu^4}{40}(5t_2 - 4t_1)\right)
\end{array}
\right]
\end{array}
\right]
\tag{7.5}
$$

4. Shortage cost per cycle in blood bank storage:

$$
BBS_{SC} = \zeta_{BBSC}\left[\int_{t_2}^{T} -\Upsilon_{BBS}(t)e^{-(r_0 + \omega_3 - 1)(t_2 + t)}\, dt\right]
$$

$$
= \dfrac{-(u_0 + \omega - 1)\zeta_{BBSC}e^{-((r_0 + \omega_3 - 1)t_2 + (\lambda_0 + \omega_1 - 1)t_1)}}{(\delta_0 + \omega_2 - 1)}
$$

$$
\left[\int_{t_2}^{T} e^{-(((r_0 + \omega_3 - 1) + (\delta_0 + \omega_2 - 1))t}\, dt - e^{-(\delta_0 + \omega_2 - 1)t_2}\int_{t_2}^{T} e^{-(r_0 + \omega_3 - 1)t}\, dt\right]
$$

$$
= \dfrac{(u_0 + \omega - 1)\zeta_{BBSC}e^{-((r_0 + \omega_3 - 1)t_2 + (\lambda_0 + \omega_1 - 1)t_1)}}{(\delta_0 + \omega_2 - 1)(r_0 + \omega_3 - 1)\left[(\delta_0 + \omega_2 - 1) + (r_0 + 1)\right]}
$$

$$
\left[
\begin{array}{l}
(\delta_0 + \omega_2 - 1)e^{-((\delta_0 + \omega_2 - 1) + (r_0 + 1))t_2} \\[1em]
+ e^{-(r_0 + 1)T}\left\{
\begin{array}{l}
(r_0 + \omega_3 - 1)e^{-(\delta_0 + \omega_2 - 1)T} \\[1em]
-\left[(\delta_0 + \omega_2 - 1)\right. \\
\left. + (r_0 + \omega_3 - 1)\right]e^{-(\delta_0 + \omega_2 - 1)t_2}
\end{array}
\right\}
\end{array}
\right]
\tag{7.6}
$$

5. Opportunity cost in blood bank storage:

$$
BBS_{LS} = \zeta_{BBLS}\int_{t_2}^{T} (u_0 + \omega - 1)(1 - e^{-(\delta_0 + \omega_2 - 1)t})e^{-(\lambda_0 + \omega_1 - 1)t_1}e^{-(r_0 + \omega_3 - 1)(t_2 + t)}\, dt
$$

$$= \frac{\zeta_{\mathrm{BBLS}}(u_0 + \omega - 1)e^{-((\lambda_0 + \omega_1 - 1)t_1 + (r_0 + \omega_3 - 1)t_2)}}{(r_0 + \omega_3 - 1)\big((\delta_0 + \omega_2 - 1) + (r_0 + \omega_3 - 1)\big)} \left[\begin{array}{l} e^{-(r_0 + \omega_3 - 1)t_2} \left\{ \begin{bmatrix} (\delta_0 + \omega_2 - 1) + \\ (r_0 + \omega_3 - 1) \end{bmatrix} \\ -(r_0 + \omega_3 - 1)e^{-(\delta_0 + \omega_2 - 1)t_2} \end{array} \right\} \\ -e^{-(r_0 + \omega_3 - 1)T} \left\{ \begin{bmatrix} (\delta_0 + \omega_2 - 1) \\ +(r_0 + \omega_3 - 1) \end{bmatrix} \\ -(r_0 + \omega_3 - 1)e^{-(\delta_0 + \omega_2 - 1)T} \end{array} \right\} \right] $$

$$\tag{7.7}$$

The total average cost per unit time of our model is obtained as follows:

$$\mathrm{BBSTC}(t_1, T) = \frac{1}{T}\big[\mathrm{BBS}_{\mathrm{OC}} + \mathrm{BBS}_{\mathrm{HC}} + \mathrm{BBS}_{\mathrm{DC}} + \mathrm{BBS}_{\mathrm{SC}} + \mathrm{BBS}_{\mathrm{LS}}\big] \tag{7.8}$$

$$= \frac{1}{T} \left[\left\{ \{\zeta_{\mathrm{BBOC}}\} + \left\{ \begin{array}{l} (\zeta_{\mathrm{BBHC}})(u_0 + \omega - 1) \\ \left[\begin{array}{l} \dfrac{t_1^2}{2} - \dfrac{(3(\lambda_0 + \omega_1 - 1) + (r_0 + 1))}{6}t_1^3 \\ + \left(\dfrac{(\theta + \omega_2 - 1)}{12} + \dfrac{(\lambda_0 + \omega_1 - 1)(r_0 + \omega_3 - 1)}{8} \right) t_1^4 \\ - \left(\dfrac{(r_0 + \omega_3 - 1)(\theta + \omega_2 - 1)}{20} - \dfrac{(\lambda_0 + \omega_1 - 1)(\theta + \omega_2 - 1)}{30} \right) t_1^5 \end{array} \right] \end{array} \right\} \right. $$

$$+ \left\{ \begin{array}{l} \zeta_{\mathrm{BBDC}}(\theta + \omega_2 - 1) \\ \left[\left(\dfrac{t_1^2}{2} - \dfrac{(r_0 + \omega_3 - 1)t_1^3}{3} \dfrac{(\theta + \omega_2 - 1)t_1^4}{8} \right) + (u_0 + \omega - 1)e^{-t_1(\lambda + r)} \right] \\ \left[\begin{array}{l} \dfrac{t_2^3}{6} - \dfrac{(r_0 + \omega_3 - 1)t_2^4}{12} + \dfrac{(\theta + \omega_2 - 1)t_2^5}{40} \\ - \dfrac{(r_0 + \omega_3 - 1)(\theta + \omega_2 - 1)t_2^6}{36} - \dfrac{t_1^2}{6}(3t_2 - 2t_1) \\ - \dfrac{(\theta + \omega_2 - 1)t_1^2}{60}(5t_2^3 - 2t_1^3) - \dfrac{(r_0 + \omega_3 - 1)t_1^3}{12}(4t_2 - 3t_1) \\ - \dfrac{(r_0 + \omega_3 - 1)(\theta + \omega_2 - 1)t_1^3}{36}(2t_2^3 - t_1^3) \\ - \dfrac{(\theta + \omega_2 - 1)\mu^4}{40}(5t_2 - 4t_1) \end{array} \right] \end{array} \right\} s \right]$$

$$\left[\begin{array}{c}\left[\begin{array}{c}\dfrac{(u_0+\omega-1)\zeta_{\text{BBSC}}e^{-((r_0+\omega_3-1)t_2+(\lambda_0+\omega_1-1)t_1)}}{(\delta_0+\omega_2-1)(r_0+\omega_3-1)\left[(\delta_0+\omega_2-1)+(r_0+1)\right]}\\[4mm]+\left\{\begin{array}{c}(\delta_0+\omega_2-1)e^{-((\delta_0+\omega_2-1)+(r_0+1))t_2}\\[2mm]+e^{-(r_0+1)T}\left\{\begin{array}{c}(r_0+\omega_3-1)e^{-(\delta_0+\omega_2-1)T}\\[2mm]-\left[\begin{array}{c}(\delta_0+\omega_2-1)\\+(r_0+\omega_3-1)\end{array}\right]e^{-(\delta_0+\omega_2-1)t_2}\end{array}\right\}\end{array}\right\}\end{array}\right.\\[20mm]\left[\begin{array}{c}\dfrac{\zeta_{\text{BBLS}}(u_0+\omega-1)e^{-((\lambda_0+\omega_1-1)t_1+(r_0+\omega_3-1)t_2)}}{(r_0+\omega_3-1)((\delta_0+\omega_2-1)+(r_0+\omega_3-1))}\\[4mm]+\left\{\begin{array}{c}e^{-(r_0+\omega_3-1)t_2}\left\{\begin{array}{c}\left[\begin{array}{c}(\delta_0+\omega_2-1)+\\(r_0+\omega_3-1)\end{array}\right]\\[2mm]-(r_0+\omega_3-1)e^{-(\delta_0+\omega_2-1)t_2}\end{array}\right\}\\[6mm]-e^{-(r_0+\omega_3-1)T}\left\{\begin{array}{c}\left[\begin{array}{c}(\delta_0+\omega_2-1)\\+(r_0+\omega_3-1)\end{array}\right]\\[2mm]-(r_0+\omega_3-1)e^{-(\delta_0+\omega_2-1)T}\end{array}\right\}\end{array}\right.\end{array}\right]\end{array}\right] \tag{7.9}$$

7.5 MULTIOBJECTIVE PARTICLE SWARM OPTIMIZATION ALGORITHM

- R: =0
- $\{M_x, N_x, U_x, V_x\}_{x=1}^{X}$:= initialize()
- for a:= 1: U
- for b:= 1: X
- for r:= 1: R
- $n_{xc}^{(a+1)} = yn_{xc}^{a} + c_1 d_1\left[V_{xc} - m_{xc}^{a}\right] + c_2 d_2\left[U_{xc} - m_{xc}^{a}\right]$
- $M_x^{a+1} = M_x^{a} + mN_x^{a} + \in^{a}$
- end
- M_x:= enforce Constraints(X)
- $Y_x := f(M_x)$
- if $M_x \nleq e\ \forall\ e \in P$
- R:= $\{e \in R/\ e \nless M_x\}$
- R:= $R \cup M_x$
- end
- end
- if $M_x \leq V_x \vee (XM_x \nless V_x \wedge V_x \nless M_x)$
- $V_x := M_x$
- end

- U_x:= select Guide(X, A)
- end

7.6 NUMERICAL ILLUSTRATION

To illustrate the model numerically, the following parameter values are considered.
$(u_0 + \omega - 1) = 110\,$units, $\zeta_{BBOC} = $ Rs. 210 per order, $(r_0 + \omega_3 - 1) = 1.15\,$unit, $(\lambda_0 + \omega_1 - 1) = 0.14\,$unit, $\zeta_{BBHC} = $ Rs. 20.10 per unit, $\theta = 0.104\,$unit, $t_1 = 0.14\,$year, $\zeta_{BBLS} = $ Rs. 8.10 per unit, $(\delta_0 + \omega_2 - 1) = 0.12\,$unit, $T = 1\,$year.

Then for the minimization of total average cost, with help of software, the optimal policy can be obtained such as $t_2 = 0.99\,$year, S = 176.9885 units, and BBSTC = Rs. 1316.28 per year.

- **PSO:** [Population] = 61, [Generations] = 601, [Cognitive learning factor] = 4.1, [Cooperative factor] = 4.2, [Social learning factor] = 1.4, [Inertial constant] = 1.5, and [Number of neighbors] = 13.

7.7 SENSITIVITY ANALYSIS

TABLE 7.1
Demand Parameter $(u_0 + \omega - 1)$

$(u_0 + \omega - 1)$	T	TCBBS
152.5	182.65	1250.53
154	182.57	1249.35
156.7	182.55	1248.23

TABLE 7.2
Demand Parameter $(\lambda_0 + \omega - 1)$

$(\lambda_0 + \omega - 1)$	T	TCBBS
10.62	132.65	1380.53
10.62	132.57	1379.35
10.7	132.55	1368.23

TABLE 7.3
Backlogging Parameter $(\delta_0 + \omega_2 - 1)$

$(\delta_0 + \omega_2 - 1)$	T	TCBBS
10.43	122.65	1898.53
10.49	122.57	1897.34
10.53	122.55	1996.23

TABLE 7.4

Selling Price (ζ_{BBLS})

ζ_{BBLS}	T	TCBBS
124.5	152.65	1478.53
124.9	152.57	1477.35
125.0	152.55	1476.23

TABLE 7.5

Deterioration Parameter ($\theta + \omega_2 - 1$)

$(\theta + \omega_2 - 1)$	T	TCBBS
10.00	172.65	1308.53
10.62	172.57	1307.35
10.7	172.55	1306.23

TABLE 7.6

Particle Swarm Optimization (PSO) Model Optimal Solution

	WW		PSO		
P	OPT	BEST	MAX	AVG	STD
1	1251.10	1271.10	1272.10	1262.10	1252.10
2	1251.11	1271.11	1272.11	1262.11	1252.11
3	1251.21	1272.21	1273.21	1263.21	1253.21
4	1251.23	1272.23	1273.23	1263.23	1253.23
5	1251.24	1272.24	1273.24	1263.24	1253.24
6	1251.54	1273.54	1274.54	1264.54	1254.54
7	1251.58	1273.58	1274.58	1264.58	1254.58

7.8 CONCLUSION

These studies incorporate some down-to-earth skin that is expected to be connected with blood bank storage inventories of any material using PSO (Tables 7.1–7.6). The occurrence of decay (deterioration) overtime and inventory reductions for any material product is a natural occurrence in real conditions using PSO. The lack of blood bank storage inventory is allowed in the model using PSO. In the luggage, customers are trained for delivery delays, and parties may be willing to wait for a short period of time to get their first choice using swarm optimization. In general, the distance end-to-end time for the next replacement is the main factor in decide whether the backlog will be accepted or not using PSO. A customer's willingness to wait for backlogging during short periods of time declines with the length of waiting time using PSO. Thus, the lack of blood bank storage inventory is allowed and partially backordered in this chapter, and the backlog rate is considered as a decreasing function of waiting time for the next replenishment using PSO. The demand rate is taken as an exponential ramp-type function of time, in which the demand decreases rapidly for some initial period and subsequently stabilizes using PSO. Since most decision-makers think that inflation does not have a significant effect on inventory policy, inflation effects are not considered in some inventory models. However, from a financial standpoint, an inventory represents a capital investment and must be completed along with other assets for the firm's limited capital funds using PSO.

REFERENCES

1. Hsieh, T.P., Dye, C.Y. and Ouyang, L.Y. (2008): Determining optimal lot size for a two-warehouse system with deterioration and shortages using net present value. *European Journal of Operational Research*, 191 (1), 180–190.
2. Benkherouf, L. (1997): A deterministic order level inventory model for deteriorating items with two storage facilities. *International Journal of Production Economics*, 48 (2), 167–175.
3. Yadav, A.S. and Swami, A. (2018): A partial backlogging production-inventory lot-size model with time-varying holding cost and Weibull deterioration. *International Journal Procurement Management*, 11 (5), 639–649.
4. Yadav, A.S. and Swami, A. (2018): Integrated supply chain model for deteriorating items with linear stock dependent demand under imprecise and inflationary environment. *International Journal Procurement Management*, 11 (6), 684.
5. Bhunia, A.K. and Maiti, M. (1998): A two-warehouse inventory model for deteriorating items with a linear trend in demand and shortages. *Journal of the Operational Research Society*, 49 (3), 287–292.
6. Goswami, A. and Chaudhuri, K.S. (1992): An economic order quantity model for items with two levels of storage for a linear trend in demand. *Journal of the Operational Research Society*, 43, 157–167.
7. Lee, C.C. (2006): Two-warehouse inventory model with deterioration under FIFO dispatching policy. *European Journal of Operational Research*, 174 (2), 861–873.
8. Sahooa, L., Bhuniab, A.K. and Kapur, P.K. (2012): Genetic algorithm based multi-objective reliability optimization in interval environment. *Computers & Industrial Engineering*, 62, 152–160.
9. Yadav, A.S. and Swami, A. (2019): A volume flexible two-warehouse model with fluctuating demand and holding cost under inflation. *International Journal Procurement Management*, 12 (4), 441.
10. Yadav, A.S. and Swami, A. (2019). An inventory model for non-instantaneous deteriorating items with variable holding cost under two-storage. *International Journal Procurement Management*, 12 (6), 690.
11. Yang, H.L. (2004): Two warehouse inventory models for deteriorating items with shortages under inflation. *European Journal of Operational Research*, 157, 344–356.
12. Zhou, Y.W. and Yang, S.L. (2005): A two-warehouse inventory model for items with stock-level-dependent demand rate. *International Journal of Production Economics*, 95 (2), 215–228.
13. Yadav, A.S., Taygi, B., Sharma, S. and Swami, A. (2017). Effect of inflation on a two-warehouse inventory model for deteriorating items with time varying demand and shortages. *International Journal Procurement Management*, 10 (6), 761.

8 Live Virtual Machine Migration Techniques in Cloud Computing

Shalu Singh and Dinesh Singh
Deenbandhu Chhotu Ram University
of Science and Technology

CONTENTS

8.1 INTRODUCTION

Over the past few decades, we have been using traditional IT system to run various application and business processes. In conventional time, IT company purchases the hardware and software that are necessary to meet the business needs. Also, companies house these technological capabilities on-site. With technological advancement, the cloud computing is evolved, which is really an economical definition of delegating the management of your IT infrastructure to the third party. Organizations that take care of the administration of this IT infrastructure are called cloud providers. In the cloud model, the resources and the applications are hosted by cloud providers, and consumers access these resources via the Internet. Live migration is an essential part of the cloud environment. It continuously requires supplementary resources [1] such as CPU cycle, memory, and network speed, which ultimately affects the execution of running applications. So, virtual machine (VM) migration operation must finish in minimum time. VM migration helps in better resource utilization and increases cloud providers' capital income. Live VM migration provides uninterrupted cloud services during migration time without compromising service-level agreement (SLA) [1] violations and achieves efficient resource utilization. Live VM migration is a practical approach in cloud computing. It has various leverages such as load [2] balancing and helps in energy reduction, better service availability, and so on.

8.1.1 VIRTUALIZATION

Virtualization is an elemental technology with the help of which one can take advantage of infrastructure-based services. It creates a protected, estimable, and segregated [1] environment for the execution of cloud applications. This technology provides the capability to a computer system to imitate the executing environment independent from the computer system itself. It is contemplation of computer resources. It allows separation of the computer operating system from its hardware. It allows a single hardware resource to act as numerous virtual resources [3] and also various hardware resources to act as a single virtual resource. It has changed the IT world in many ways. It provides increased performance in the high-end side of the computer market where supercomputers give boundless compute power that helps in furnishing thousands of VMs. When there is a lack of space to accommodate thousands of physical machines, then virtualization benefits. Also, it provides the extra computing power needed by large-scale applications.

This chapter is constituted as follows. Section 8.2 gives an outline of the literature work. The VM migration mechanism is explained in Section 8.3. The various VM migration techniques used in cloud computing are provided in Section 8.4, which is followed by a conclusion in Section 8.5.

8.2 LITERATURE WORK

Numerous research articles have been published pertaining to different techniques of VM migration in cloud computing. A few research articles have been mentioned to give a glimpse of the same in the following section.

The author [4] has proposed energy-aware resource utilization model. This algorithm aids in lowering the energy utilization of cloud infrastructure by using server consolidation and also does not sacrifice the performance of cloud applications. This technique maximizes energy efficiency with the efficient utilization of cloud resources. It also provides a high speed of convergence.

Firefly optimization (FFO) is a bio-inspired algorithm and is used for VM migration problem. This algorithm is chosen for its fast convergence speed and comprehensive optimization attainment. FFO-based VM migration saves space for storage using the network-attached storage (NAS) [5]. With the help of NAS, [5] one can save images of VM and data, which ultimately saves space and provides better data access capabilities. It also helps in solving the energy consumption problem.

The author [6] has proposed a hybrid technique to reduce energy consumption in cloud data centers. This technique uses a combination of an artificial bee colony and bat algorithm. The implemented technique was compared with performance parameters such as success and failure rate [6] and energy consumption. The proposed technique is implemented in CloudSim simulator, and it achieved the minimum energy consumption and failure rate [6] with the maximum success rate.

The author [7] has proposed a method for VM migration that ponders the quality of service requirement so that energy consumption and SLA violations were reduced to some extent. The proposed work focuses on load balancing [7]–based mechanism. The work was implemented in CloudSim toolkit and compared with an existing technique like nonpower-aware, single-threshold policies. The empirical results show that the proposed technique reduces power consumption to a greater extent.

The author [8] has designed the smart elastic scheduling algorithm that dynamically distributes the load to physical hosts to obtain a load-balanced system [8]. The proposed algorithm reduces power consumption and processing time of jobs. It also minimizes the number of migrations [8] and energy consumption and helps in preventing performance degradation while maintaining the load of the entire system.

8.3 VIRTUAL MACHINE MIGRATION

VMs achieved elasticity and resource utilization. The resources in VMs can be allocated and deallocated as per demand and availability. There is a possibility that there are not enough resources available for the current demand at a given time, more especially in the case of private clouds. In these cases, to get on-demand VMs (typically) from the public cloud, the cloud bursting technique is used. Therefore, there is a transparent way of extending the computing capacity for the user. VM migration is the most dominant method that gives data center operators the capacity to accept the placement of VMs to get high performance and reduces energy consumption [9] and improves resource utilization.

8.3.1 VIRTUAL MACHINE MIGRATION MECHANISM

A large-scale cloud data center, which may contain various varied physical nodes, forms the infrastructure for the VM migration process. Every node has its own CPU which can be multicore; RAM and its network bandwidth help the nodes for their characterization. The user requests for the allocation of various VMs, with the help of MIPS, network bandwidth, and the amount of RAM. The violation of SLA happens in the case where VM does not get the requested amount of resource. In the architecture of the software system, there are dispatcher, local and global managers, and virtual machine monitor (VMM). Local managers monitor the current utilization of the node's resources. The local managers help in selecting the VMs that can be moved from one host to another host in the following cases:

1. When resources are 100% utilized then there is a risk of SLA violation which requires migration of VMs from one host to another host.
2. One VM has exhaust network due to high data transfer because it is communicated with another VM that is assigned to another physical host.
3. When the temperature crosses some margin in order to cool down the nodes, VMs have to be migrated to another host. The information about the utilization of resources has sent to the global managers by the local manager, and VMs are selected for the actual migration process.

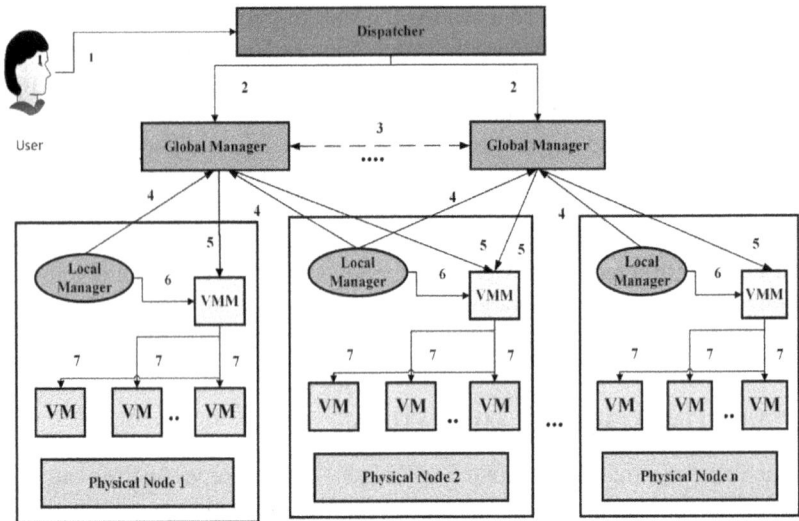

FIGURE 8.1   Virtual machine (VM) migration (VMM) mechanism.

The VM migration process consists of the following steps as shown in Figure 8.1:

1. Requests for new VM provisioning.
2. Discharging the requests for VM provisioning that helps in allocating the requests.
3. Global managers share information about current resource utilization and the VMs that have to be allocated on physical machines.
4. The local manager then passes the report about resource usage and the VMs that are needed to be migrated to the respective global managers [10].
5. The global managers then give commands to VMM that helps in optimizing current allocation.
6. Local managers check their hosts and give commands for VM resizing.
7. After receiving commands from local managers, VMM assists in executing the migration of VMs as well as resource management.

8.4 MIGRATION TECHNIQUES

There are three types of live VM migration techniques: pre-copy VM migration, post-copy VM migration, and hybrid live VM migration as shown in Figure 8.2. Each technique is discussed in detail in the following sections.

8.4.1 PRE-COPY VIRTUAL MACHINE MIGRATION

In pre-copy VMM, when the migration process starts, a destination server is chosen for the VMM. Meanwhile, resource reservation is done at the server site. It has two phases: iterative and stop-and-copy.

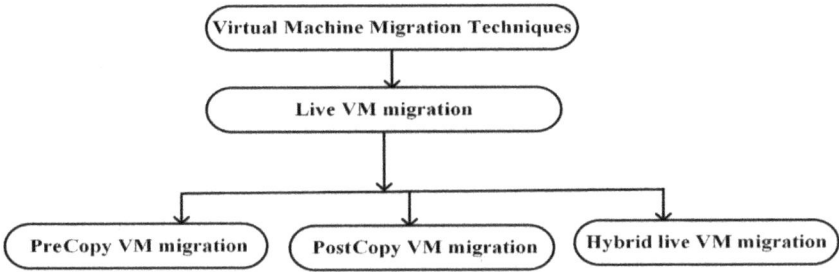

FIGURE 8.2   Live VM migration techniques. VM, virtual machine.

1. Iterative phase: In this phase, because of the repetitive operation, whole VM memory is assumed to be dirty. The dirty pages are repetitively sent to destination server during migration. All the dirty pages of VMs are transmitted regularly while VM is running on the source server.
2. Stop-and-copy phase: During this phase, the migration controller determines and transfers remaining dirty pages to the destination server with the VM states such as CPU state, VM memory, and so on as shown in Figure 8.3. After that, VM is continued at the destination server, and migration process ends.

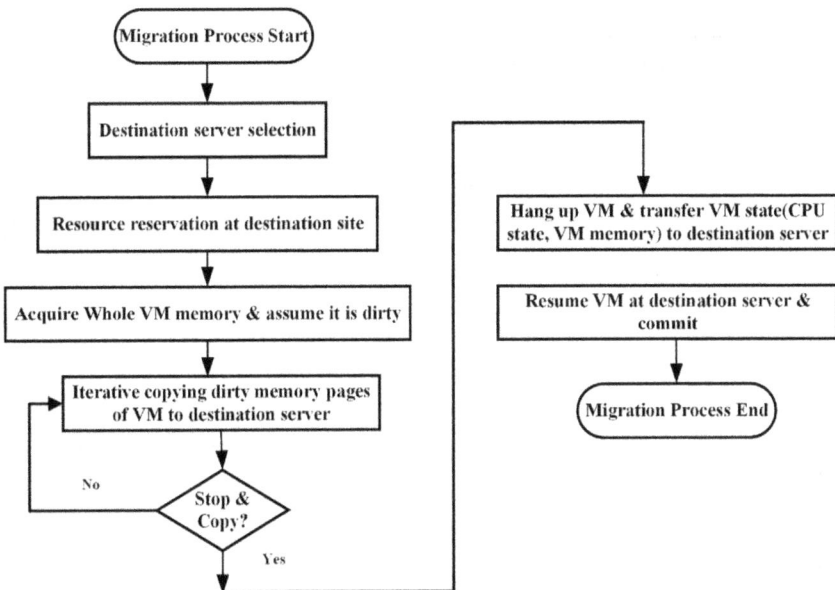

FIGURE 8.3   Flowchart of precopy VM migration. VM, virtual machine.

8.4.2   POST-COPY VIRTUAL MACHINE MIGRATION

In the post-copy VM migration method, VM minimum states such as CPU, registers, and I/O are transferred before the VM starts at the destination server. Then, VM is continued at destination server with the transmitted state as shown in Figure 8.4.

After that, dirty memory pages are actively pushed from source to destination server; if any page fault occurs during the transmission, then meanwhile faulty pages are copied from the source server. When all the VM pages are transmitted successfully, then the migration process ends.

8.4.3   HYBRID LIVE VIRTUAL MACHINE MIGRATION

It incorporates the merits of both the pre-copy and post-copy [1, 11] VMM techniques that help in improving total migration time and service downtime [11]. In Figure 8.5, hybrid live VMM is described in five phases:

1. During the first phase, resource reservation is done at the destination server.
2. During the second phase, it determines and transmits the VM working set to the destination server [1, 11].
3. In the third phase, VM minimum state [1, 11] is transmitted to the receiver server.

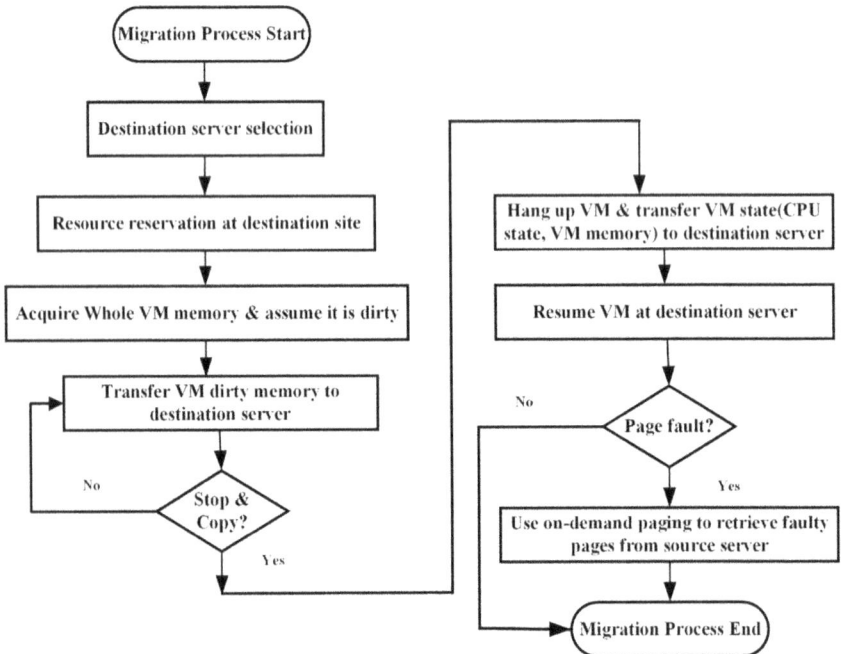

FIGURE 8.4   Flowchart of postcopy VM migration. VM, virtual machine.

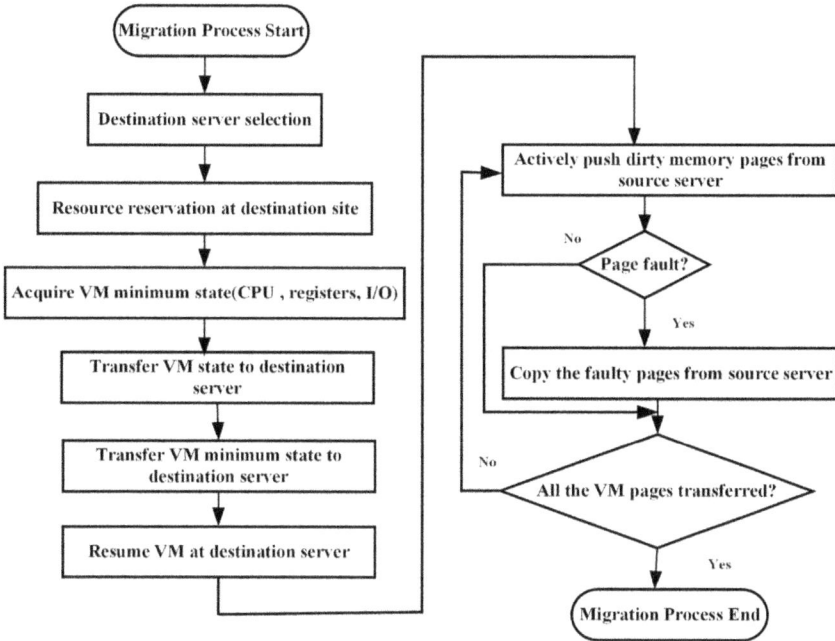

FIGURE 8.5 Flowchart of hybrid live VM migration. VM, virtual machine.

4. During the fourth phase, VM is continued at the receiver server.
5. In the last phase, if page faults occur, then demand paging is used to retrieve the faulty pages from source server, and if there is no page fault, then migration process ends directly.

8.5 CONCLUSION

Cloud computing has reshaped the world of the IT industry. It has been visioning earlier that cloud computing will bestow the services of fundamental computing. Cloud computing offers various resources that can be accessed anywhere through the Internet. Live VMM is a mechanism of shifting the VMs from one physical machine to another while applications are running on them. The jobs that are being executed on the VMs must be accessible all the time to cloud users. This is only viable when migration is done with zero downtime. Live VM migrations help in achieving system maintenance, load balancing, energy management, and resource sharing. The authors have described virtualization, energy efficiency, and VMM mechanism. We have also discussed pre-copy, post-copy, and hybrid live VM migration techniques.

ACKNOWLEDGMENTS

We would like to thank University Grant Commission, Government of India, for providing financial aid under National Research Fellowship grant no. NF-2017-18/31287.

REFERENCES

1. Ahmad, R., Gani, A.: Virtual machine migration in cloud data centers: a review. Taxonomy, and open research issues. *The Journal of Supercomputing* 71(7), 2473–2515 (2015).
2. Li, Y., Lan, Z. (2004) A Survey of Load Balancing in Grid Computing. In: Zhang, J., He, J.H., Fu, Y. (eds) *Computational and Information Science*. CIS 2004. Lecture Notes in Computer Science, vol 3314. Springer, Berlin, Heidelberg.
3. Zhan, Z.H., Zhang, G.Y., Ying-Lin, Gong, Y.J., Zhang, J. (2014) Load Balance Aware Genetic Algorithm for Task Scheduling in Cloud Computing. In: Dick, G. et al. (eds) *Simulated Evolution and Learning*. SEAL 2014. Lecture Notes in Computer Science, vol 8886. Springer, Cham.
4. Kansal, N., Chana, I.: Artificial bee colony-based energy-aware resource utilization technique for cloud computing. *Concurrency and Computation: Practice and Experience* 27, 1207–1225 (2014).
5. Kansal, N., Chana, I: Energy-aware virtual machine migration for cloud computing - A firefly optimization approach. *Journal of Grid Computing* 14, 327–345 (2016).
6. Karthikeyan, K., Sunder, R, Shankar, K, Lakshmanaprabu, S.K., Vijayakumar, V, Mohamed, E, Gunasekaran M Energy consumption analysis of Virtual Machine migration in cloud using hybrid swarm optimization (ABC–BA). *The Journal of Supercomputing* 76, 3374–3390 (2018).
7. Kamran, N. B.: QoS-aware VM placement and migration for hybrid cloud infrastructure. *The Journal of Supercomputing* 74, 4623–4646 (2018).
8. Nashaat, H., Nesma, A. Rizk, R: Smart elastic scheduling algorithm for virtual machine migration in cloud computing. *The Journal of Supercomputing* 75, 1–24 (2019).
9. Min, J., Zeng, Y., Zhang, P., Gao, G. (2013) The Design and Implementation of Load Balancing System for Server Cluster. In: Yang, G. (eds) *Proceedings of the 2012 International Conference on Communication, Electronics and Automation Engineering*. Advances in Intelligent Systems and Computing, vol 181. Springer, Berlin, Heidelberg.
10. Beloglazov, A., Abawajy, J., Buyya, R.: Energy-aware resource allocation heuristics for efficient management of data centers for Cloud computing. *Future Generation Computer Systems* 28, 55–68 (2012).
11. Choudhary, A., Govil, M. C., Singh, G., Awasthi, L. K., Pilli1, E. S. and Kapil, D.: A critical survey of live virtual machine migration techniques. *Journal of Cloud Computing: Advances, Systems and Applications* 6, 23 (2017).

9 Dynamic Exchange Buffer Switching and Blocking Control in Wireless Sensor Networks

R. Ganesh Babu
SRM TRP Engineering College, Tiruchirappalli, TN, India

M.N. Saravana Kumar and R. Jayakumar
Erode Sengunthar Engineering College, Erode, TN, India

CONTENTS

9.1 INTRODUCTION

The hubs focusing on remote sensor systems are sorted out in an unfriendly condition, and hubs have the capacity to detect physical marvels [1]. The sensor hubs have low battery force, and they have physical zone and have no source reestablishing their capacity. So a vitality-productive convention structure is needed through which the existence time of the system is improved. As of late, the sensor systems have numerous applications such as well-being checking, ecological observing, industry robotization, and so on [2]. Enormous measure of information and ceaseless streams required a greater number of assets than the accessible; thus, clog emerges in the system [3]. The clog causes parcel delays, lining postponement, and hubs will become associated with poor control of overhead [4]. Dealing with these conditions in a smooth way, clog control is a significant task. The control strategies are utilized to control the clog in the system, for example, lessening the traffic rate and viable assets of the executives [5]. The proposed convention DABCC is a widespread correspondence model for dealing with both huge just as consistent progression of information from the source to sink. This convention relies upon the rest of the vitality, staying cradle, and the sensor hub. This is the most intended part for viable cradle on the board in the wireless multimedia sensor networks (WMSNs).

9.2   RELATED WORKS

Blocking control in the WSN is a significant observation; however, it is a troublesome activity to distinguish where clog happens in the system. Numerous strategies are centered to control the clog by decreasing information pace of the source or asset control ideas or mixes of both [6]. They considered the line removes just as direct settings in each individual advanced neighbor just as recognizing the bottleneck in each hub by methods for new plans. By utilizing the estimation of bottleneck, each hub unexpectedly alters its bundle communicationspeed notwithstanding relatedload coordinating between them to stay away from bottleneck just as connected issues. Direness-based case precisely grouped resistor plan to distinguish in bottleneck system by methods for line-up plan of the hubs [7]. The plan is for the most part applied in fire disclosure and home automation notwithstanding interrelated uses [8], in which bottleneck focuses are recognized other source similarly suggest that once the hubs and bottleneck. Here each showing up bundle is requested as exceptional desperation in addition to bringing down need parcels. During clog, low need parcels are disposed.

9.3   PERFORMANCE EVALUATION AND THE METHODS

9.3.1   Trust-Based Blocking Control Protocol

The proposed protocol is designed in this section for universal communication model which can be implemented in WMSNs for both large and continuous data flow. The first subsection describes a creation of static hierarchical topology for initial deployment of the sensor nodes. The next subsection defines a community confidence evaluation. In the next part, constructing a safe packet using DES algorithm [9] is presentedto transfer a secure packet based on the CEM to next hop. The estimates on cost function are shown in the next section. By using different congestion metrics, congestion is identified, and the last segment focuses primarily on the implementation of an effective congestion reduction technique [10].

9.3.2   Trust Evaluation

In WSN's neighboring hubs, it is implied that the hubs are under the radio transmission scope [11]. The hub trust is a subordinate substance of the period, and it may shift depending on the exhibition of point in exchange. This history assesses the pace of confidence of the matter just as the suggestion of the neighboring hub. Here, trust metrics (information and control package delivery, parcel postponement) and QoS (quality of service) characteristics were considered in appreciation of the hub's behavior. The level of confidence taken out since the certitude known as primary trust (PT), and secondary trust (ST) are segregated from the neighboring hubs' proposal. Such important ones are described in Figure 9.1 just as the auxiliary confidence assessments. Actually, the faith of hub Q depends on its own understanding (PT). To determine the optional confidence (ST) of hub Q, hub P assembles hub Q data from its neighbors R, S, T, U, and V.

Guidelines: The nodes P and Q are direct trust calculated based on various trust metrics as follows:

$$P,Q = \sum_{m=1}^{k} W_m \times t \ m_m^{P,Q} \qquad (9.1)$$

where $tm_m^{P,Q}$ is metric of the node P and node Q.

$$TP,Q = WD \times PT \ P,Q + \ WI \times ITP,Q \qquad (9.2)$$

where WD and WI are the primary and secondary trust loads, respectively. Estimated confidence parameters are included in the neighboring table.

9.3.3 Securing the Data Packet Generation

WSN is utilized in different submissions because of the innovation improvement of the remote correspondence, for example, unwavering quality, adaptability, modest, and simple to send to anyplace. WSNs are effectively undermined by inactive

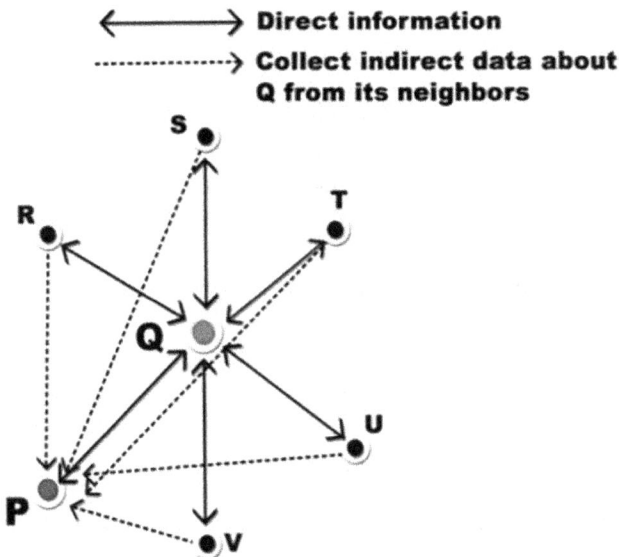

FIGURE 9.1 Nodes between the trust relationships.

listening stealthily and dynamic interruption [12]. Inactive listening implies that the programmer makes some private data and those are embedded alongside unique message, whereas in the dynamic interruption, the programmer erases chosen data in the message or supplements adulterated data alongside it or else imitationhubs end the well-being record just as nonnotoriety of WSN. To keep away from this sort of assaults, part of cryptographic systems are created. The triple DES (TDES) calculation is utilized in the proposed technique, in which DES is applied multiple times. Three sorts of TDES modes are proposed, for example,

DES-EEE3: Three-layered DES encryptions by methods for threeself-administering keys.

DES-EDE3: Three DES forms in request encode follow with unscramble again finished by scramble self-deciding keys.

DES-EEE2 just as DES-EDE2: Similar to the prior game plans barring that the head in addition to third assignments practices the comparable key.

Among the three, the proposed technique utilizes the second idea, for example, encode and decode scramble with three autonomous keys K1, K2, and K3. Each key is 64 bits since quite a while ago, planned for a total distance of the key 192 bits. The TDES turns triple level lesser than the DES. In any case, it offers additional security when accurately utilized. The procedure planned for decoding the correspondence is comparable as the encryption; barring the keys stays actualized in turned-around request. The encryption and decryption processes are spoken to underneath:

$$CT = (EK3(DK2(EK1(PT)))) \text{ (Encryption operation)}$$
$$PT = (DK1 (EK2 (DK3 (CT)))) \text{ (Decryption operation)}$$

(9.3)

The key size of DES stays 64-piece long; anyway true key utilization through DES is 56 piece extensive. The littlest critical piece in each byte is applied in the interest of equality bit. It is fixed so that an odd quantity of ones remains constantly in each byte. If we overlooked the bits of equality in each byte, the subsequent key length is 56 pieces in DES and 168 bits in TDES. The charge estimation manner (CEM) is built on the leftover cradle, extraordinary vitality just as the confidence estimates. The following table is used to properly discover the CEM during package transmission:

$$CEM(Q) = \max\{1 / d \ (\ TQ, P + RB(P) + RE(P)\}$$

(9.4)

Alot of sensor hubs in a particular level fulfill high outstanding vitality and staying cushion with least separation; then, one hub is picked sequentially as essential CEM, which goes about as next bounce for steering. This next process of determining a bounce proceeded until the sink was found. Earlier hubs are pronounced as extra CEM, and its subtleties in the neighboring hub are refreshed in Figure 9.2.

The difference between P and Q is determined according to the following:

$$d = \sqrt{(N_{Q.x} - N_{P.x})^2 + (N_{Q.y} - N_{P.y})^2}$$

(9.5)

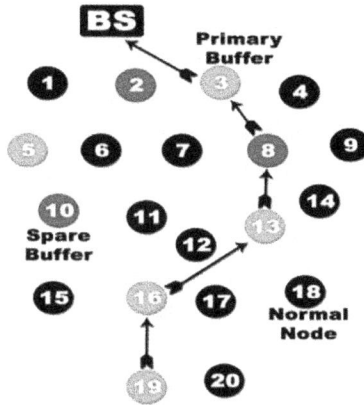

FIGURE 9.2 Buffer switching nodes.

Typically, organized administrations utilize twofold plans, open space trench conspire just as radio-based plan. If the message separation "d" is smaller than the starting stage d_0 separation, open space channel conspire is currently being implemented, or multipathway blurring strategy is being applied. The radio program characterizes the vitality used for transmitting k-bit information by separating d, if the vitality utilized is characterized with the radio plan in Figure 9.3.

The transmitting energy is computed as follows:

$$\text{ETX} = \text{Eele}(ki, di) + \text{Eap}(ki,\ di)$$

$$= ki \times \text{Eele} + ki \times \text{fs}\ (d < d0) \tag{9.6}$$

$$= ki \times \text{Eele} + ki \times \text{mp}\ (d >= d0)$$

The receiving energy is computed as follows:

$$\text{ERX} = ki \times \text{Eele} \tag{9.7}$$

The hub's remaining energy is the proportion of the present vitality to the full sensor point vitality in Figure 9.4. The vitality of the present situation is determined as

$$\text{EP} = \text{Ei} - \left\{ \text{ETX}(k,d) + \text{ERX}(k) \right\} \tag{9.8}$$

where Ei is the underlying vitality of the sensor hub (assessed during various leveled creation),

ETX(k, d) is the energy used to move a k number of bits to separate d, and
ERX(k) is the energy used to obtain k bits.
In sensor hub, whole vitality is characterized as follows:

$$\text{ET} = \text{ETX}(k,d) + \text{ERX}(k) + \text{Eproc} + \text{Esense} \tag{9.9}$$

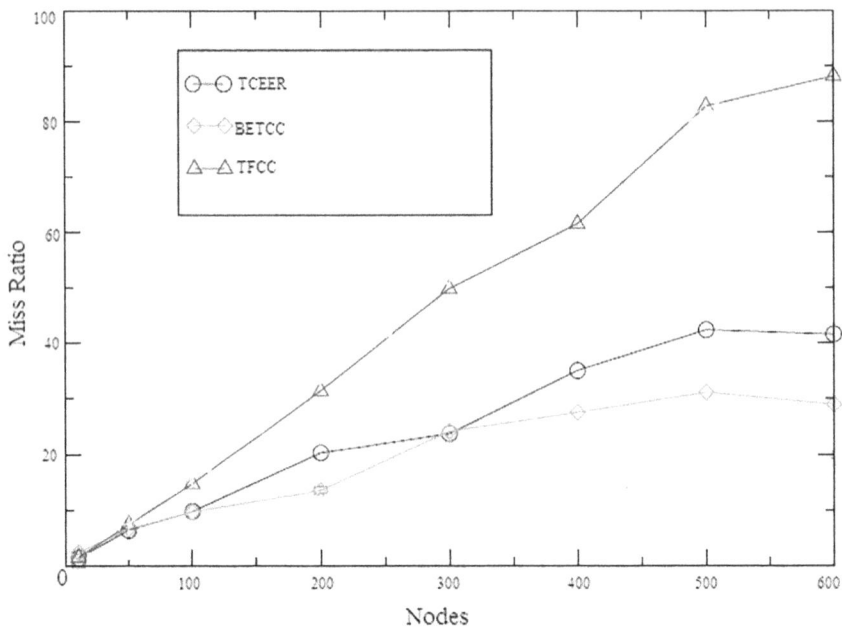

FIGURE 9.3 Nodes versus miss ratio.

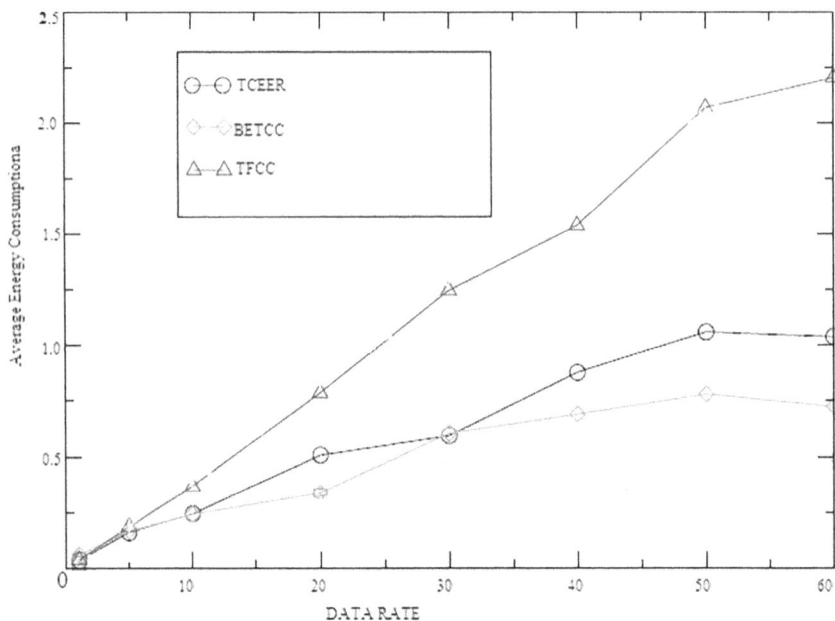

FIGURE 9.4 Data rate versus average energy consumption.

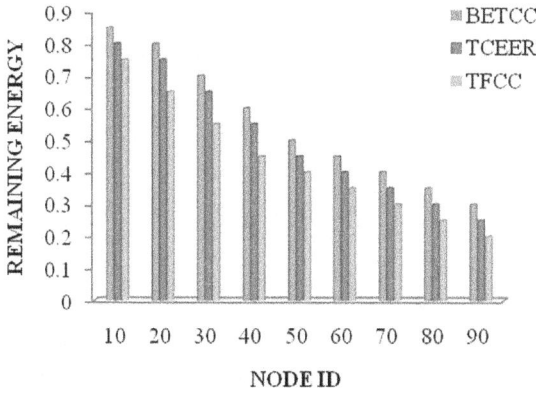

FIGURE 9.5   Remaining energy of the nodes.

The energy remaining in hub P is calculated as follows:

$$RE(P) = EP/ET \qquad (9.10)$$

where EP is the present vitality – the sensor hub resides, and ET is the whole vitality of the sensor hub.

The remaining cushion of the node is processed dependent on present cradle position just as the anticipated measure of parcels to be conveyed beginning from hub P toward its most extreme level in Figure 9.5. The remaining support of the hub P is characterized in condition,

$$RP(P) = Bi - \sum_{n=1}^{N} K_N \qquad (9.11)$$

where Bi is the current supporting status of the sensor hub, K_N is the number of bundles, N transmitted from P to its greater level.

9.4   CONCLUSION AND FUTURE WORK

The proposed convention is designed to rely on, remaining in WMSNs, buffer, excellent power, and trust-based congestion control to maintain a strategic distance from clog. Here, various static leveled topologies for introductory WSN are made. The confidence estimate of the neighboring hubs is evaluated using trust measurements. The safe information bundle is made by utilizing TDES; this information is sent to the following bounce. The following bounce choice depends on the CEM. This strategy sorts the cradle into essential and extra support. On the off chance that there is no blockage in the system, essential support is dealt with the information parcel. During information transmission, blockage happened implies that the cradle exchanging is performed. The essential support made a wake-up signal through which the extra cushion to deal with blocked locale is empowered successfully. The productive next bounce determination of the proposed convention improves the rest of the vitality level of the cradle and diminishes the number of bundle errors.

REFERENCES

1. Kafi, M.A., Djenouri, D., Ben-Othman, J., Badache, N.: Congestion Control Protocols in Wireless Sensor Networks: A Survey. *IEEE Communications Surveys and Tutorials* 16(3), 1369–1390 (2014).
2. Ganesh Babu, R.: Helium's Orbit Internet of Things (IoT) Space. *International Journal of Computer Science and Wireless* 1(1), 123–124 (2017).
3. Sonmez, C., Incel, O.D., Isik, S., Donmez, M.Y., Ersoy, C.: Fuzzy-Based Congestion Control for Wireless Multimedia Sensor Networks. *EURASIP Journal on Wireless Communications and Networking* 1, 1–17 (2014).
4. Ganesh Babu, R.: Mismatch Correction of Analog to Digital Converter in Digital Communication Receiver. *International Journal of Advanced Research Trends in Engineering and Technology* 3(19), 264–268 (2016).
5. Sakthidevi, I., Srievidhyajanani, E.: Secured Fuzzy Based Routing Framework for Dynamic Wireless Sensor Networks. In: *International Conference on Circuits, Power and Computing Technologies*, pp. 1041–1046. Nagercoil, India (2013).
6. Ganesh Babu, R., Amudha, V.: Cluster Technique Based Channel Sensing in Cognitive Radio Networks. *International Journal of Pure and Applied Mathematics* 119(16), 3341–3354 (2018).
7. Bakra, B.A., Lilien, L.T.: Extending Lifetime of Wireless Sensor Networks by Management of Spare Nodes. *Second International Workshop on Communications and Sensor Networks* 34(1), 493–498 (2014).
8. Ganesh Babu, R.: WIMAX Capacity Enchancements Introducing Full Frequency Reuse Using MIMO Techniques. *International Journal of Advanced Research in Biology Engineering Science and Technology* 2(16), 1–7 (2016).
9. Kafi, M.A., Djenouri, D., Ben-Othman, J., Ouadjaout, A., Badache, N.: Congestion Detection Strategies in Wireless sensor Networks: A Comparative Study with Test Bed Experiments. *5th International Conference on Emerging Ubiquitous Systems and Pervasive Networks* 37, 168–175 (2014).
10. Ganesh Babu, R.: Cognitive Radios Spectrum Allocation in Wireless Mesh Networks. *International Journal of Global Research and Development* 1(6), 289–294 (2016).
11. Ganesh Babu, R., Karthika, P., Manikandan, G.: Polynomial Equation Based Localization and Recognition Intelligent Vehicles Axis using Wireless Sensor in MANET. In: Second International Conference on Computational Intelligence and Data Science in association with Elsevier-Procedia Computer Science 167, 1281–1290 (2020).
12. Karthika, P., Ganesh Babu, R., Jayaram, K.: Biometric Based on Steganography Image Security in Wireless Sensor Networks. In: Second International Conference on Computational Intelligence and Data Science in association with Elsevier-Procedia Computer Science 167, 1291–1299 (2020).

Index

For Product Safety Concerns and Information please contact our EU
representative GPSR@taylorandfrancis.com
Taylor & Francis Verlag GmbH, Kaufingerstraße 24, 80331 München, Germany

www.ingramcontent.com/pod-product-compliance
Lightning Source LLC
Chambersburg PA
CBHW070738220326
41598CB00024BA/3462